Scientifica 7

For Key Stage 3 Science

Peter Ellis • Phil Godding • Derek McMonagle
Louise Petheram • Lawrie Ryan
David Sang • Jane Taylor

Published in 2004 by:
Nelson Thornes Ltd
Delta Place
27 Bath Road
CHELTENHAM
GL53 7TH
United Kingdom

04 05 06 07 / 10 9 8 7 6 5 4 3 2 1

A catalogue record for this book is available from the British Library

ISBN 0 7487 7981 7

Illustrations by Mark Draisey, Ian West, Bede Illustration
Page make-up by Wearset Ltd

Printed and bound in Spain by Graficas Estella

CONTENTS

SCIENTIFICA

See which lesson you are studying

This shows what you should hope to learn in this lesson. If you don't understand these things at the end, read through the pages again, and don't be afraid to ask teacher!

LEARN ABOUT
How Scientifica works

You can use the things you learn in Science in other subjects too. These panels will help you watch out for things that will help you in other lessons like Maths, Geography and Citizenship. Sometimes they will contain handy hints about other sections of the book.

CHALLENGE

It's really important to develop good computer skills at school. These ICT Challenges will provide lots of interesting activities that help you practice.

Welcome to Scientifica

Why should Science textbooks be boring? We think Science is amazing, and that's why we've packed this book full of great ideas. You'll find tons of amazing facts, gruesome details, clever activities and funny cartoons. There's lots of brilliant Science too!

Here are some of the main features in Scientifica. There are lots more to discover if you look…

Get stuck in

Whenever you see a blue-coloured panel, it means it's time to start doing some science activities. This blue panel provides a set of simple instructions you can follow. Your teacher may also have a sheet to help you and for you to write on.

Meet the Scientifica crew!

Throughout the book you may see lots of questions, with four possible answers. Only one is correct. The answers **are in different colours**. If the teacher gives the class coloured cards to vote with, it will be easy to show your vote.

Did you really understand what you just read?

Are you sure?

Won't these questions help you check?

SUMMARY QUESTIONS

At the end of each lesson, there is a set of questions to see if you understood everything.

☆ See those stars at the beginning of each question?

They tell you whether a question is supposed to be Easy (☆), Medium (☆☆) or Hard (☆☆☆).

Freaky insects, expanding bridges, boiling hot super stars... These are the most fantastic facts you can find!

Find out how scientists worked out what we know so far. Don't worry! There's plenty more for the scientists of the future to find out.

Gruesome science

Lethal clouds of poison gas, dead human skin cells, killer electric eels... Sometimes Science can be just plain nasty! Why not learn about that too?

UNIT REVIEW

There are loads of homework questions at the end of each Unit. There are lots of different types too. If you complete all the SAT-style questions, the teacher may be able to tell you what Level you are working at. Do your best to improve as you go along!

DANGER! AVOID THESE COMMON ERRORS

'Er, the Sun goes out like a light-bulb at night, right?' People make mistakes about science all the time. Before you leave the topic, this will help you make sure you're not one of them!

Phenomenal performance

If you do *brilliantly* in the lesson, your *extraordinary* teacher may ask you to turn towards the back of this *fantastic* book. There are lots of *super* activities for you to try in the *Phenomenal Performance* section.

Key words

amazing
brilliant
phenomenal

Keywords are a handy way of remembering a topic. Some might be scrambled up though!

We think you'll enjoy Scientifica, and hopefully Science too. Best of luck with your studies!

The (other) Scientifica crew – *Lawrie, David and Jane*

7A

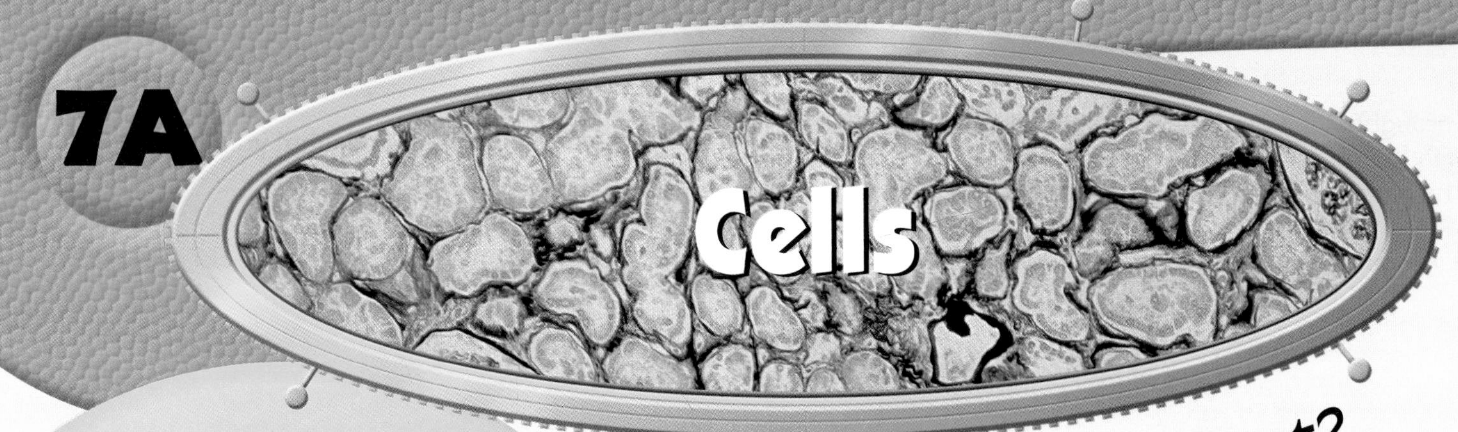

Coming soon

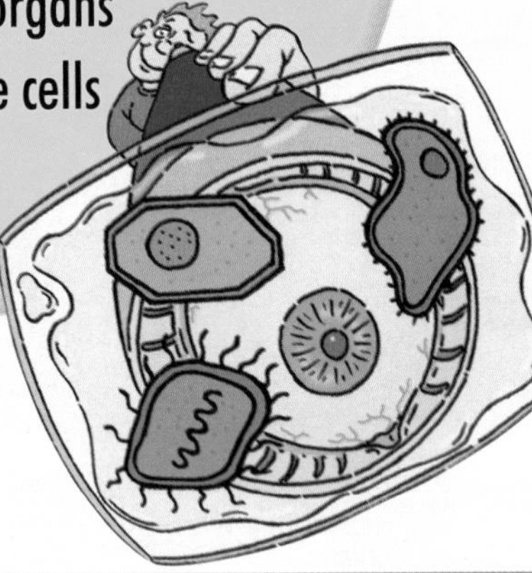

What's it all about?

How will we know if there is life on Jupiter's moons? We will look for signs of life. All living things carry out life processes.

What are living things made of? Sugar and spice and all things nice? No. Muscles and bones? Getting closer. All living things are made of **cells**.

In this unit we will look at life processes, cells and microscopes.

What do you remember?

You already know:

- that we need to breathe and digest food
- about a plant's roots, stems and leaves
- what the heart, skeleton and muscles do

1 All living things carry out activities they need to stay alive. Make a list of these activities. The first two have been done for you.

Living things reproduce
Living things feed

2 Match the part to its job.

part: heart root lungs leaf
job: takes in water from the ground pumps blood makes food using light absorbs oxygen from the air

3 Look at the picture. It is something familiar magnified using a microscope. Can you identify it?

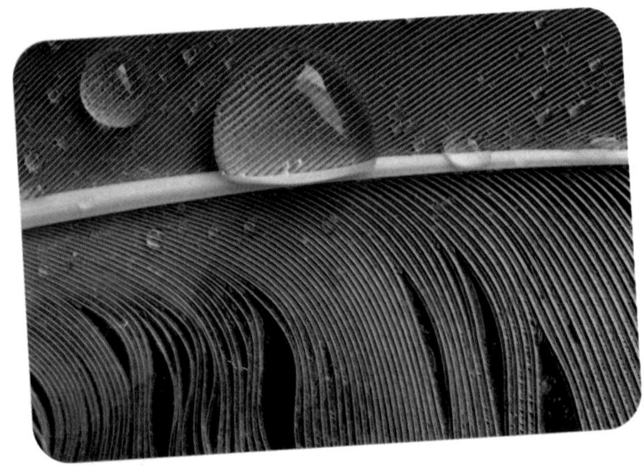

What is this?

Ideas about cells

QUESTIONS

The students at Scientifica are about to find out about cells.

- Why can't Mike, Molly, Pip or Benson see any cells?
- Is Reese correct – is the liver an organ? Name three organs in the human body and give their jobs.
- Molly is worried that her cells are dying. Can you explain where new cells come from?

7A1 Microscopes

LEARN ABOUT
- using microscopes
- observing a specimen

Too small to see

Even if we look carefully, we cannot see what skin is made of. This is because our eyes cannot see anything smaller than about one-tenth of a millimetre. Skin cells are a hundred times smaller so we have to magnify them to see them.

We look at very small things with a **microscope**. A light microscope uses two lenses to magnify the view of a small sample of material. The sample is called a **specimen**. You can see the parts of a light microscope in the picture.

The view that you see is called the **image**. It is magnified as it passes through the first lens. It is magnified again as it passes through the second lens.

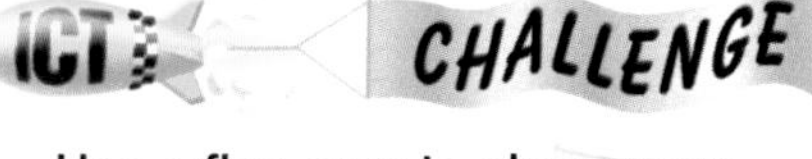

Use a flex-cam to show your slide to the rest of your group. Tell them about what you can see.

Lenses change the direction of the rays of light.

Humans are made of at least 50 million million cells.

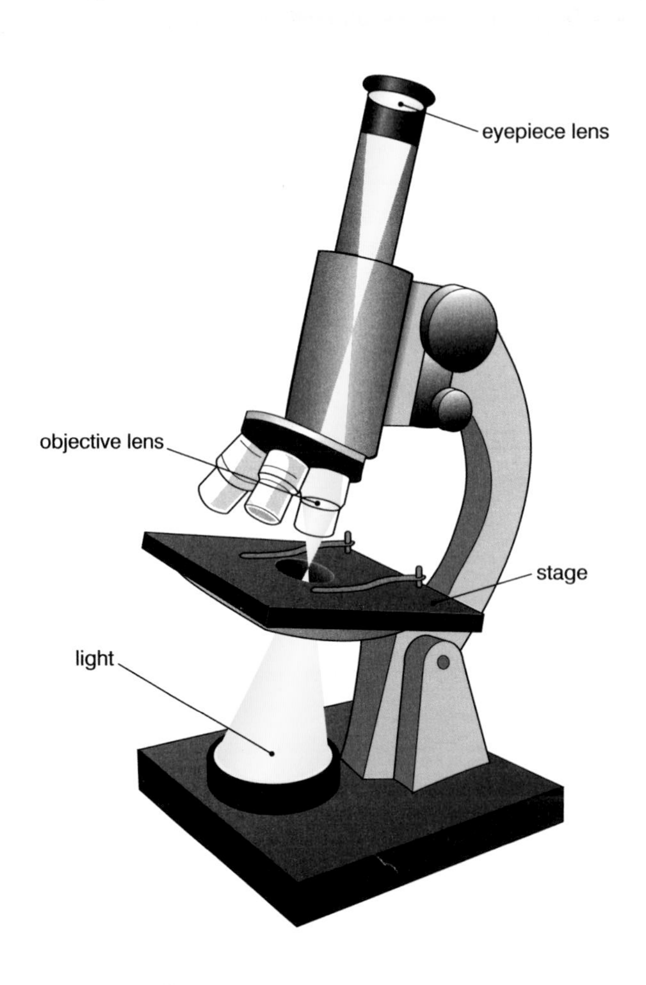

The parts of a microscope

Seeing cells

If we want to see cells we must cut the specimen into very thin slices. The slices are called **sections**. These are thin enough for light to pass through. One section is put on a microscope **slide** with a drop of water. It is then covered with a **cover slip**. It is now ready to go on the microscope.

The photo shows a section of a plant stem as it looks through a microscope. You can see that it has hundreds of small, rounded shapes packed together. These are the cells.

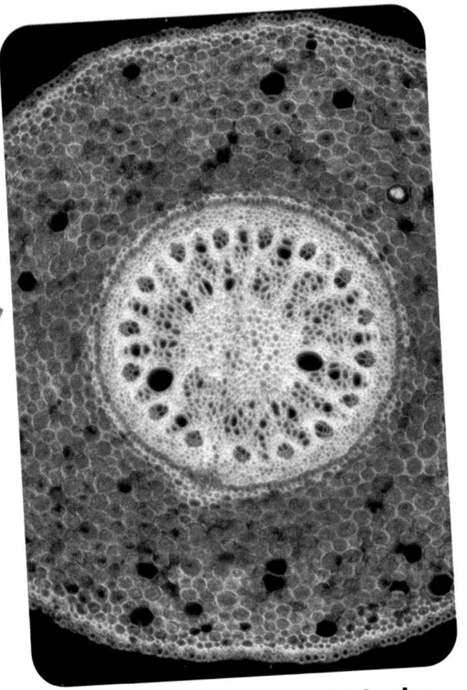

How many different kinds of cell can you see in the plant stem?

Many plant cells are shaped like bricks. When the cells are cut across, they look rectangular or roughly circular.

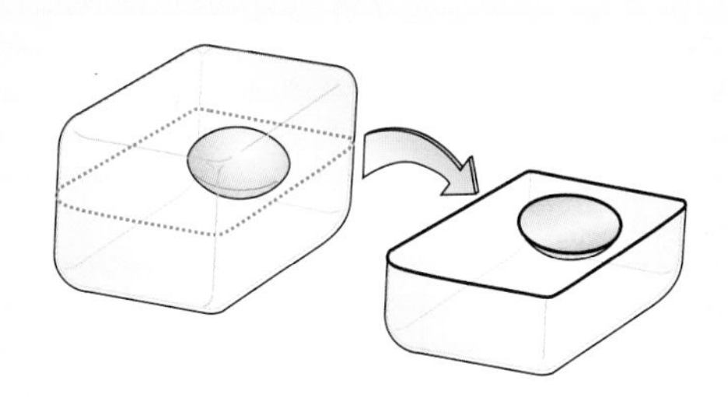

A cell shaped like a cube looks like a square when you see it in a section

Cells are almost transparent. Scientists have developed **stains**, which dye substances in a cell. These make it easier to see structures inside a cell.

Would you like another section of toast?

Using a microscope

- Set up your microscope. Practise focusing so that you can see a specimen on a slide clearly.
- Look at a clear plastic ruler under the microscope.
  a) Can you see the millimetre markers?
  b) How far is it from one side of the image to the other?
- Use your microscope to look at prepared slides of plants and animals.

SUMMARY QUESTION

1 ☆ Explain what each of the following is:
a) a slide
b) a cover slip
c) a section
d) a stain

Key words
cover slip
image
microscope
section
slide
specimen
stain

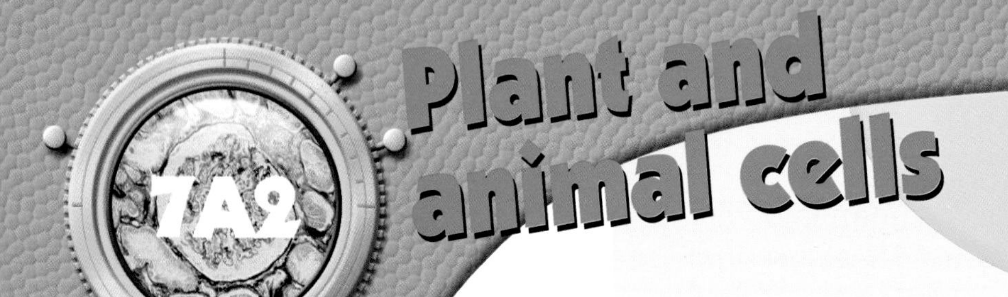

Plant and animal cells

LEARN ABOUT
- what is life
- what is in a cell
- plant and animal cells

How do we know what is alive?

It is hard to believe that barnacles on a rock are alive. How do we know? We know because they carry out all the life processes. If something does not do **all** these processes, then it isn't alive.

Life process	Animals	Plants
movement	move around from place to place	parts of the plant move
detect changes	use senses and nerves to detect changes	detect light, day length, and which way is up
breathe	take in oxygen and produce carbon dioxide	use oxygen to release energy from foods make food with carbon dioxide
grow	get larger	get larger
reproduce	make young versions of themselves	make seeds and spores
produce wastes	make urine and release carbon dioxide	release oxygen into the air
take in energy	eat foods	use light to make foods

Cells

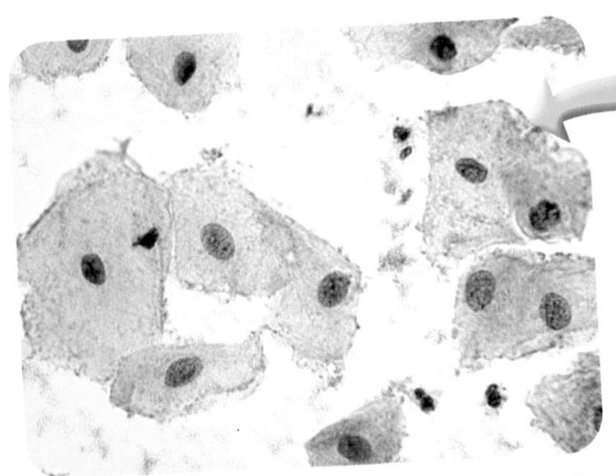

These cells come from the inside of someone's mouth

Cells are the smallest building blocks of animals and plants. A thin **cell membrane** encloses the cell's contents. It controls substances entering or leaving the cell.

Cells are filled with **cytoplasm**. Cytoplasm is a jelly-like substance. There are many useful chemicals dissolved in it.

Cells have a large **nucleus** in the cytoplasm. The nucleus directs the cell's activity. It is sometimes described as the 'control centre'. The nucleus has a set of instructions for carrying out activities in the form of **genes**.

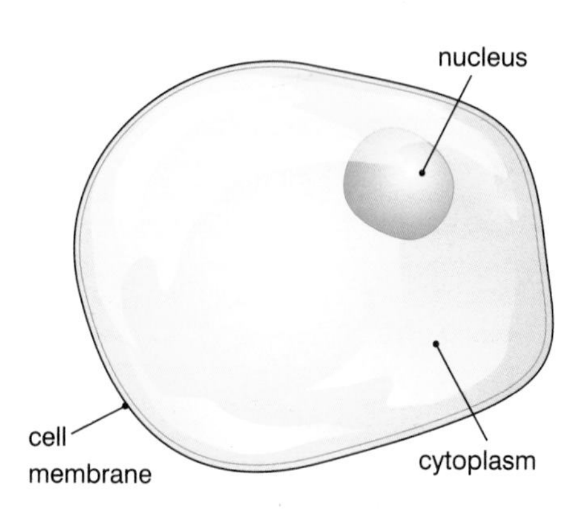

Animal cell

Plant cells

Plant cells also have:

- a **cell wall**. This is a layer outside the cell membrane. The cell wall is made of cellulose. It is a tough substance and keeps the cell's shape.
- **chloroplasts** in the cytoplasm. These give plants their green colour. Chloroplasts use light energy to make food for the plant.
- a large bubble of water called a **vacuole** inside the cell. There are many useful substances dissolved in the water.

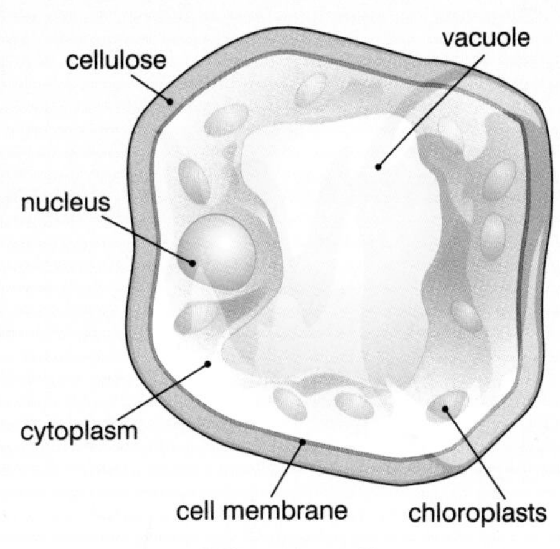

Plant cell

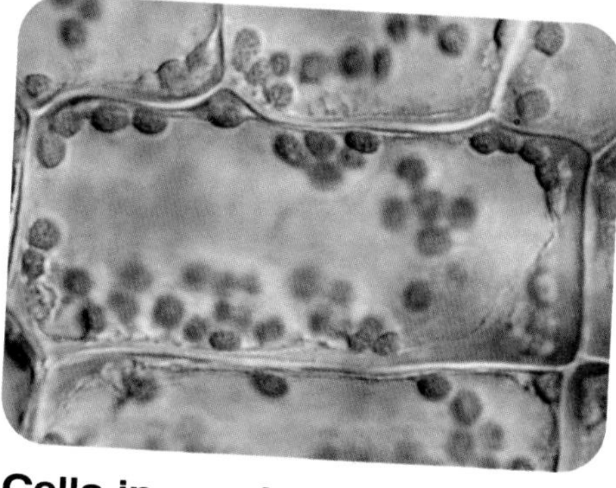
Cells in pondweed are packed with chloroplasts

Q1 Which part of a plant cell is made of cellulose?

nucleus
cell membrane
vacuole cell wall

Looking at cells

- Take a small piece of the layer of cells lining the layers of an onion. Prepare a section and use a microscope to look at the cells.
- Look for the cell wall and cytoplasm.
- Carefully draw a few cells. Label the cell wall and nucleus.

SUMMARY QUESTIONS

1 ☆ A car takes in petrol, produces gases from the exhaust and signals when it is going to turn. Why isn't it alive?

2 ☆ Look at the picture.
a) What are A B, C and D.
b) Which is made mainly of water?
c) Which makes a plant cell appear green?

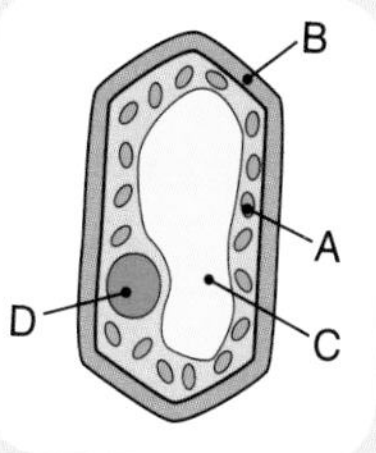

3 ☆☆ Using the headings 'Animal and plant cells' and 'Only in plant cells', sort the following into two groups:

nucleus cell wall chloroplast
cytoplasm vacuole cell membrane

Key words

cell membrane
cell wall
chloroplast
cytoplasm
genes
nucleus
vacuole

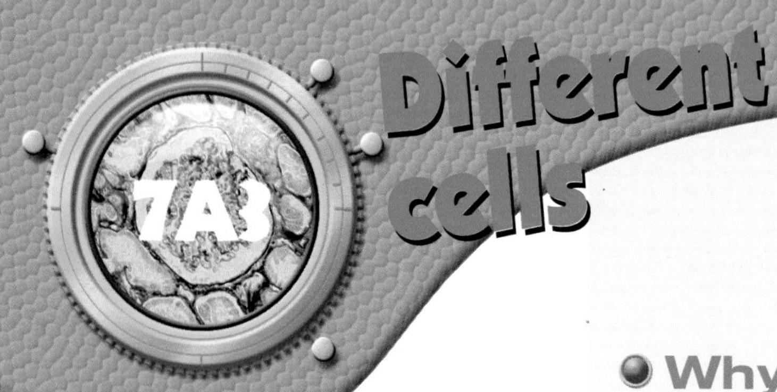

Learn about

- different sorts of cells

Why do we need different cells?

Would you use a small car to deliver a big double bed? No. It's the wrong shape. Just as we use different vehicles for different jobs, so our body uses different cells for different jobs.

Specialised cells

Animals and plants have specialised cells. These cells have some extra features to help them do their job. For example, there are cells in the eye that detect light. They contain a special chemical that changes when light hits it.

Q1 Look at the pictures. Explain how each cell's features help do its job.

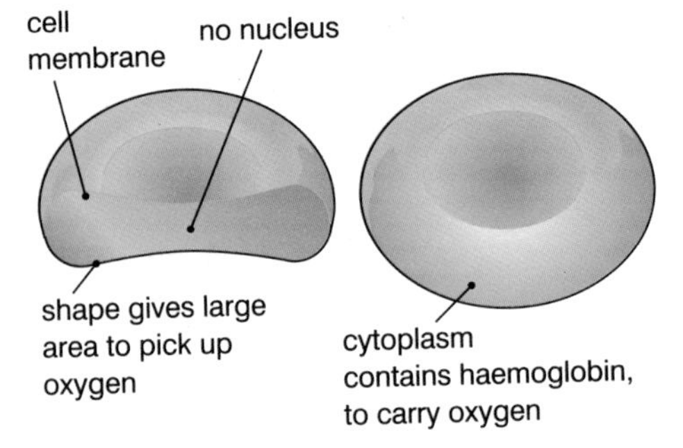

Red blood cells carry oxygen to every part of the body

Root hair cells absorb water and minerals from the soil

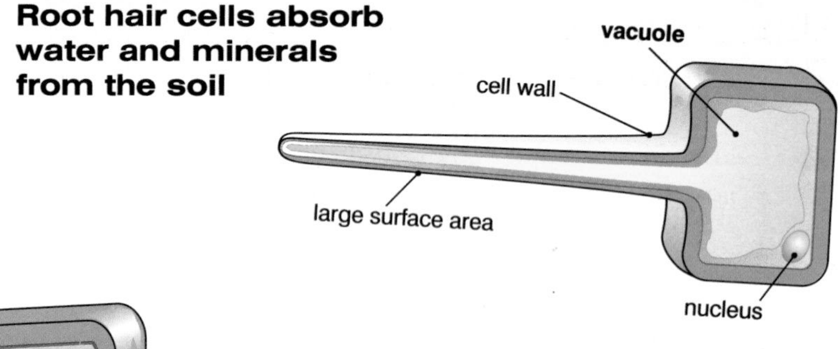

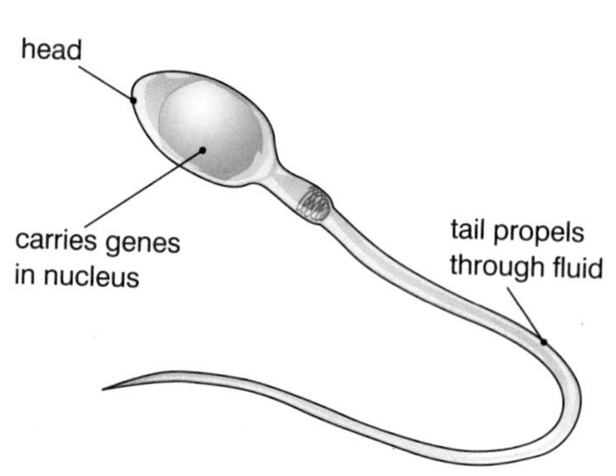

A sperm cell. Its job is to deliver a nucleus, containing genes, to an egg.

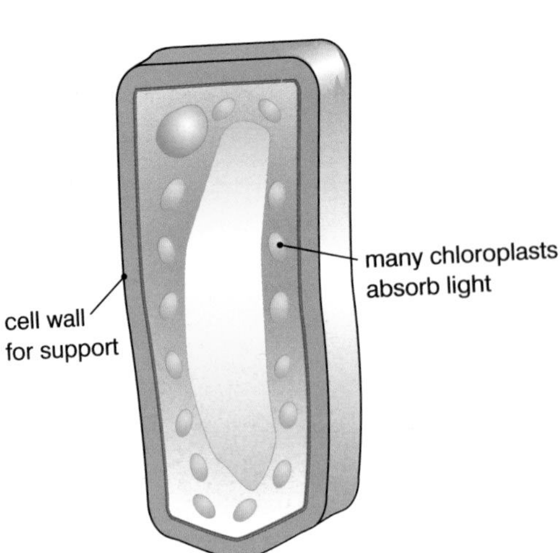

Palisade cells use light to make sugars for the plant

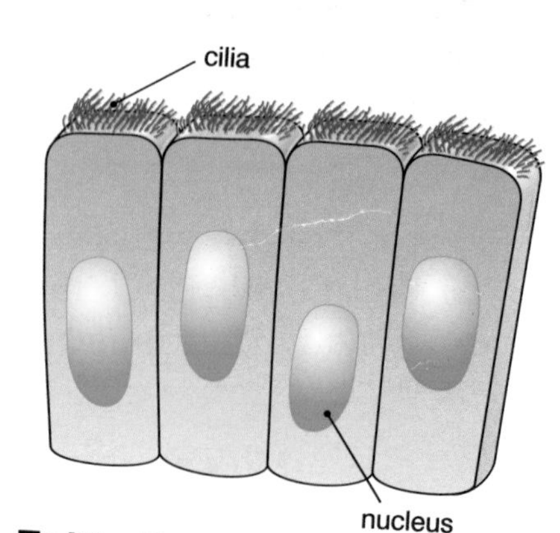

Epithelium cells line some of the tubes in the body. The small 'hairs' beat, moving substances along the tube.

Looking at specialised cells

Potato plants make starch. They store it in cells in the tuber.

- Put a very thin slice of potato and two drops of iodine stain on a microscope slide. Iodine makes starch grains turn blue-black.
- Look at the starch grains. Estimate how many are there in a cell.
- Draw two or three cells with starch grains. What do you think the potato uses starch for?
- Make a model of one of the cells you have learned about.

Gruesome science

House dust is mainly old skin cells from you and your pets.

Blood transfusions

People can lose a lot of blood when they have an operation or a nasty accident, or give birth. They are given a blood transfusion to make up the missing blood.

People usually have whole blood in a transfusion. However, some patients need just red blood cells or just plasma. Plasma is the liquid part.

Blood donors give about half a litre of blood each time. This contains over 2 billion blood cells.

SUMMARY QUESTIONS

1 ☆ Choose a cell for each job from the list:
job: **making food** **moving mucus along the windpipe** **absorbing water** **carry oxygen**
cell: red blood cell root hair cell palisade cell epithelium cell

2 ☆☆ Look at the cells in the picture.
a) Which cell features can you see?
b) Is it a plant or an animal cell?
c) Which feature helped you decide?

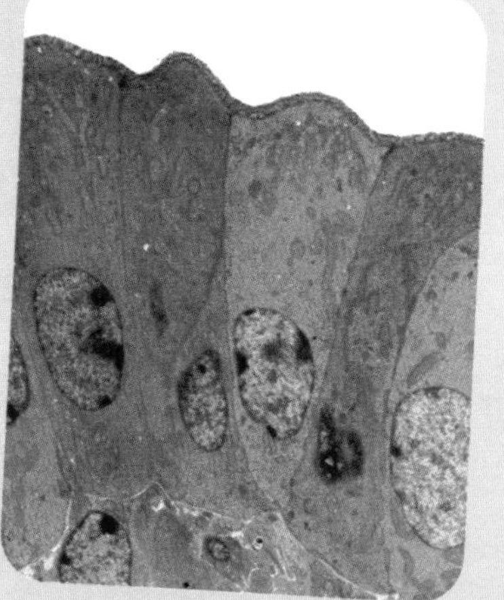

Is this a plant or an animal cell?

Key words

epithelium cell
palisade cell
red blood cell
root hair cell
sperm cell

7A4 Tissues and organs

LEARN ABOUT
- tissues
- organs
- where the organs are

Working together

You use lots of different parts of your body when you have a PE lesson. They work together so that you can move about smoothly.

For example, when you play football or netball, you use your leg muscles. Leg muscles need oxygen to work. They need your lungs to absorb oxygen. You need blood to carry it from your lungs, and your heart to pump your blood to your muscles.

Organs

Organs are the parts of your body that carry out life processes, such as breathing and reproduction. Usually several organs work together on one process. Your lungs and heart are organs involved in getting oxygen to your leg muscles.

Leg muscles also need energy. They need sugar, which comes from the food you eat. Several more organs help you to digest and absorb food.

Q1 Your liver is an organ involved in digesting food. How many others can you think of?

You can see how organs are linked together in our body to carry out our life processes in the pictures opposite.

Your skin is one huge organ – as big as a duvet cover!

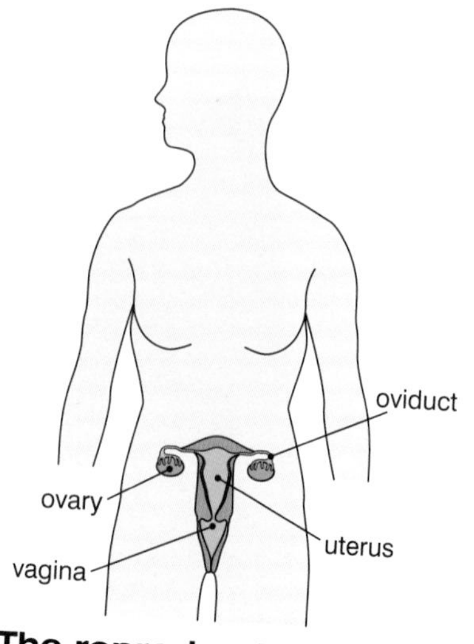

The reproductive system

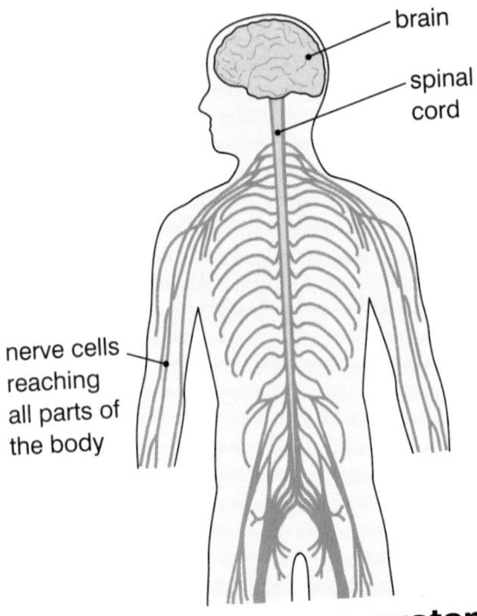

The nervous system

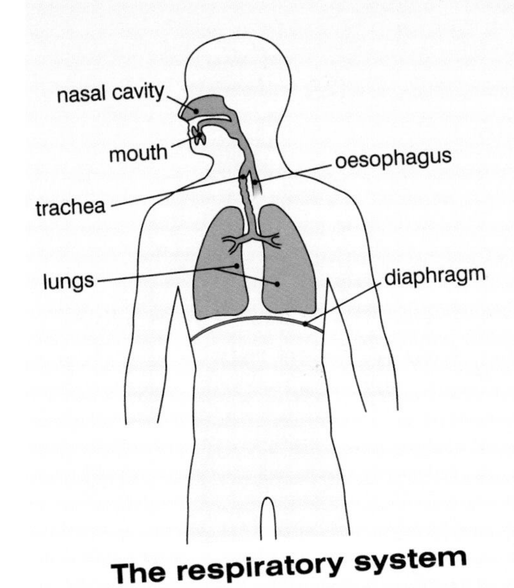

The respiratory system

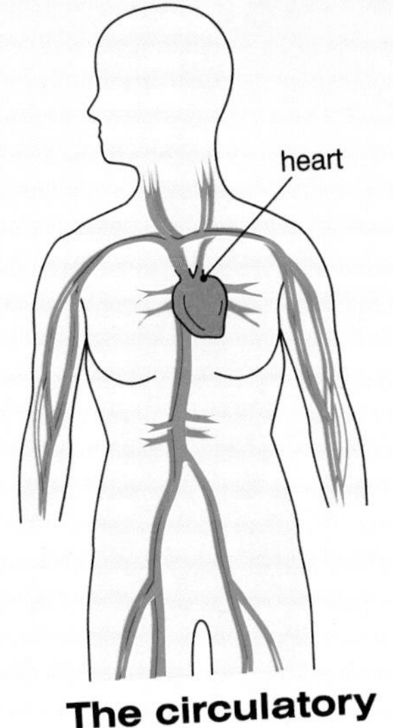

The circulatory system

mouth
oesophagus
liver
stomach
pancreas
intestines

The digestive system

Tissues

Cells in the body are grouped together into **tissues**. A tissue is a group of similar cells carrying out the same job.

There are many different tissues in an organ. In the lungs, while one sort of cell keeps the airways clear of dust, another sort absorbs oxygen.

Muscle is also a tissue. It is a cluster of muscle cells that work together to pull on bones.

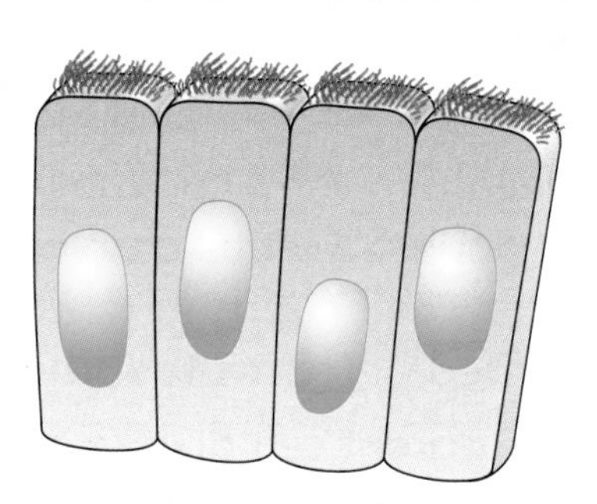
A tissue is a layer of similar cells doing a job, like these cells lining your airway

Finding out about organs

- Make a map of the human organs used in breathing, digestion and reproduction.

SUMMARY QUESTIONS

1 ☆ Read 'working together'. Identify one organ and one tissue.

2 ☆ List the parts of the body we use to breathe.

3 ☆☆ The picture shows some of the organs in a human body. Identify each organ and explain what it does.

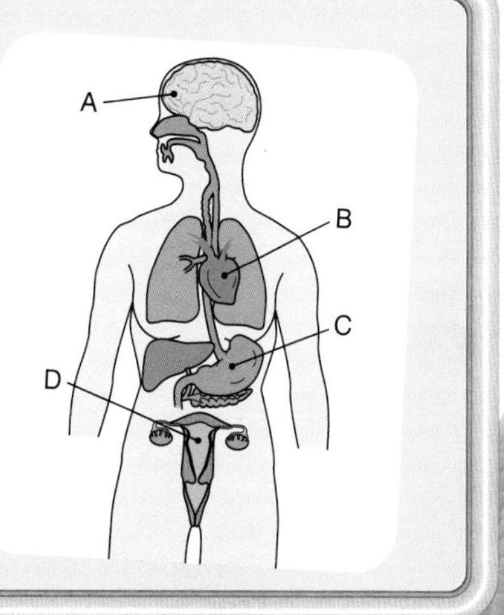

Key words

organ
reproductive system
tissue

7A5 Making more cells

LEARN ABOUT
- how cells make more cells

Why do we need new cells?

It is amazing that new skin cells heal cuts in just a few days. We make new cells all the time. Some are to repair damage – for example, to replace lost blood. We also make more cells when we grow.

Making more cells

New cells are made by **cell division**. A parent cell divides to make two smaller cells. These new cells grow larger and develop their specialist features.

First the parent cell makes a copy of all its genes. Genes carry information that a new cell needs to do its job. They are carried on **chromosomes**.

The parent cell divides all its contents evenly into two halves.

The two halves separate to become two new 'daughter' cells. Each daughter cell now has a copy of the parent's genes. The new cells are exactly the same as the cell they came from.

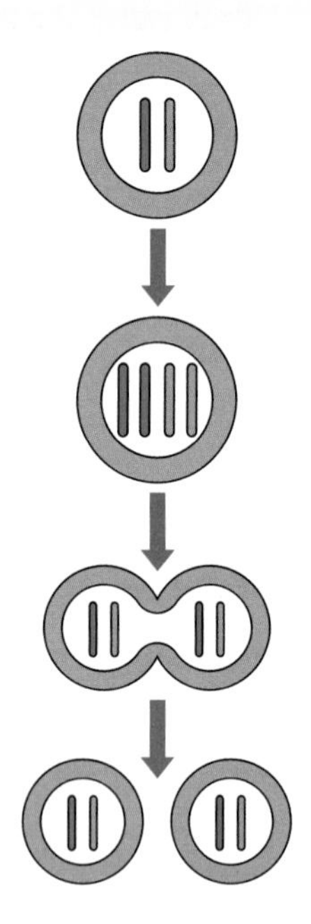

How cells make more cells

Q1 If you start at 9 a.m. with one bacterium that can divide every 20 minutes, estimate how many will you have by 3.30 p.m. Choose from:

500 5000 50 000 550 000

You make about 25 million new red blood cells every day to replace old ones.

Asexual reproduction

Bacteria and other single-celled animals and plants found in ponds and damp places, have only *one* cell. They also reproduce by copying their genes. They share the cell contents between two daughter cells. This is called **binary fission**.

Bacteria can reproduce very quickly in a warm, moist place with plenty of food. Reproduction which does not need two parents is called **asexual reproduction**.

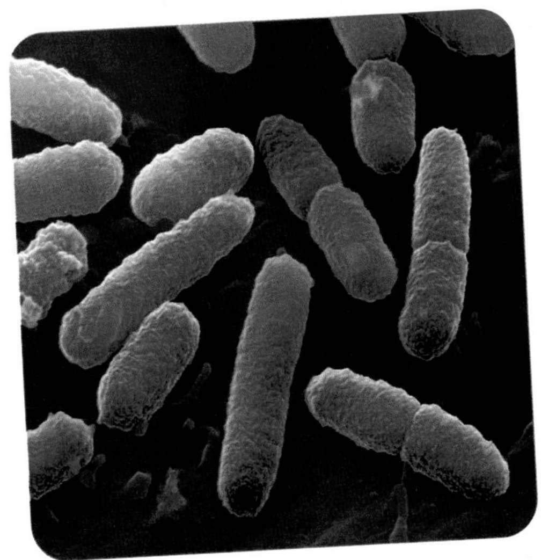

Bacteria reproducing. You can see the 'waist' where two new cells are separating.

Strawberry reproduction

Strawberry plants make seeds in the usual way, but they can also reproduce by asexual reproduction. They grow shoots, called runners, along the ground.

First the cells at the tip of the runner divide rapidly. The new cells develop into roots and leaves. Within a few days they are a new plant.

It is identical to its parent because it is made from its parent's cells and carries the same genes.

The strawberries on the new plants will be every bit as tasty!

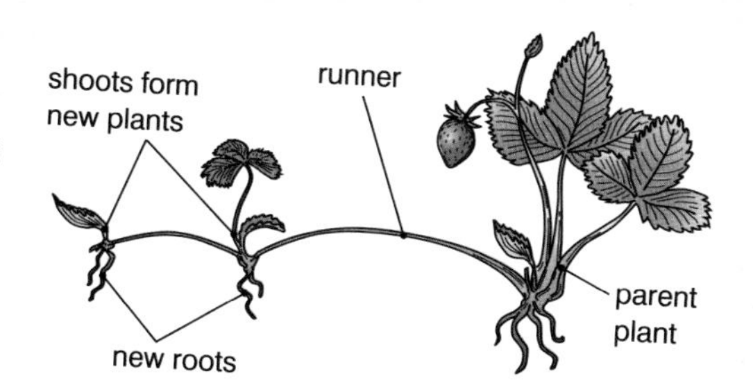

Making more strawberry plants

Looking at dividing cells

- Use a microscope to look at dividing cells in a root.
  a) Where are the smallest cells?
  b) How do the cells change as they get older and the root grows down?

SUMMARY QUESTIONS

1 ☆ Define cell division, binary fission, asexual reproduction.

2 ☆☆ Garlic, daffodil and tulip plants all grow from a bulb in the ground. At the end of the year there are tiny little bulbs growing around the original bulb. Do you think the new small bulbs will be the same as the parent or different? Why?

3 ☆☆ Genes are important for a strawberry's taste. Explain why the strawberries on the new plants will taste the same as those on the parent plant.

Key words

asexual reproduction

binary fission

cell division

chromosomes

7A6 Pollen and ovules

LEARN ABOUT
- pollen and ovules

Flowers

Flowers are a plant's reproductive organs. The anthers on the top of the stamens make pollen. **Ovules** are made in the ovary.

Pollen is carried from one plant to another by insects or by the wind. This is called **pollination**. Pollination is the first step in the process of making seeds. Pollen and ovules contain specialised cells that will eventually produce a new plant.

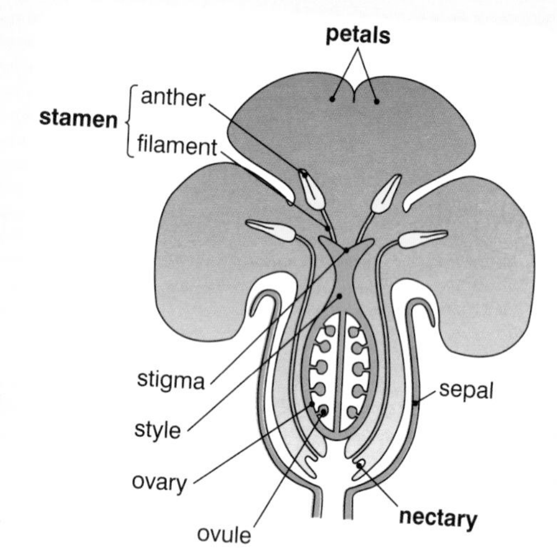

Flower structure

Bees visit brightly coloured flowers to collect nectar and pollen

Pollen

Pollen grains have a hard, protective coat around the outside. Specialised cells inside the grain deliver a nucleus to an ovule.

Pollen grains are adapted for the journey from one plant to another. The tough coat stops them drying out. Pollen grains carried by the wind are very light so that they can float in air. The hooks and spikes on pollen grains carried by insects catch in their hairs.

Q1 Do you think the pollen grains in the picture are transferred by insects or wind?

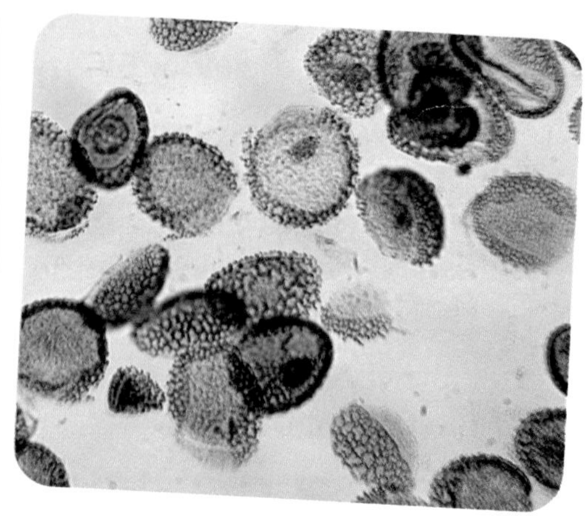

Pollen

Ovules

Some of the cells inside the ovule store food for the seed when it starts to grow. Other cells make the egg nucleus that is needed to make the new plant.

Fertilisation

Newly arrived pollen is trapped in a sticky, sugary solution on the top of the stigma. Each pollen grain grows a long tube that pushes down between the cells to the ovary.

The pollen tube then pushes into an ovule. It releases a nucleus, which combines with the egg nucleus in the ovule. This is called **fertilisation**. After fertilisation the ovule can develop into a seed.

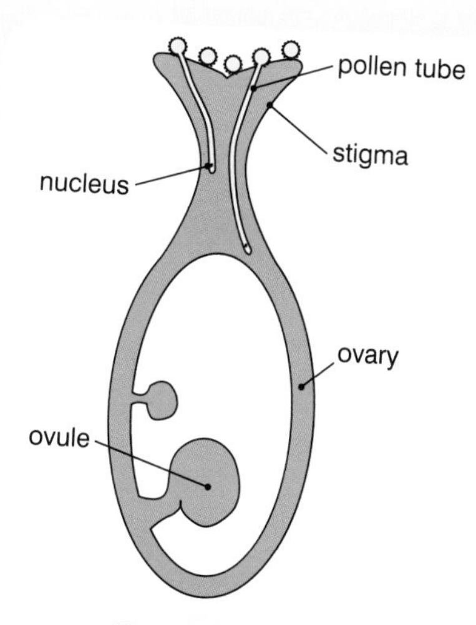

Fertilisation

ICT CHALLENGE

Search on the Internet for pictures of pollen grains. Compare pollen spread by wind with pollen carried by an insect.

Looking at pollen

- Look at pollen grains with a microscope.
  Can you see anything that would help them to be carried by the wind or an insect?
- Look at the flower your pollen came from.
  Roughly how far would the pollen tube have to grow so as to reach the ovule from the stigma?
- Grow pollen grains in a sugary solution. Use a microscope to compare how two different sorts of pollen grow.

SUMMARY QUESTIONS

1 ☆ What is the difference between pollination and fertilisation?

2 ☆☆ Make a flow chart of how plants reproduce. You can include information you have learned before.

Key words

fertilisation
ovules
pollen
pollination

IDEAS AND EVIDENCE

The first microscopes

Antony van Leeuwenhoek was born in Holland in 1632. He knew about Robert Hooke's microscope work observing animal and plant materials. Hooke gave us the word 'cell' for the small units he could see in his specimens. Hooke's simple microscope magnified objects by about 25 times.

Leeuwenhoek developed microscopes that could magnify over 200 times. They had clearer and brighter images than before. He knew that he needed good lighting to see anything clearly. He used his microscopes to discover a world of new living things.

Leeuwenhoek worked in a different way from other scientists. He was willing to change his ideas when his **observations** did not fit with his predictions. He also developed new ways of investigating living things. Some of these are part of the scientific method we use today.

Leeuwenhoek discovered bacteria, single-celled pond creatures, sperm cells and red blood cells. He wrote careful descriptions of his observations, and he used an illustrator for the drawings. These drawings are so good that anyone today could recognise his specimens.

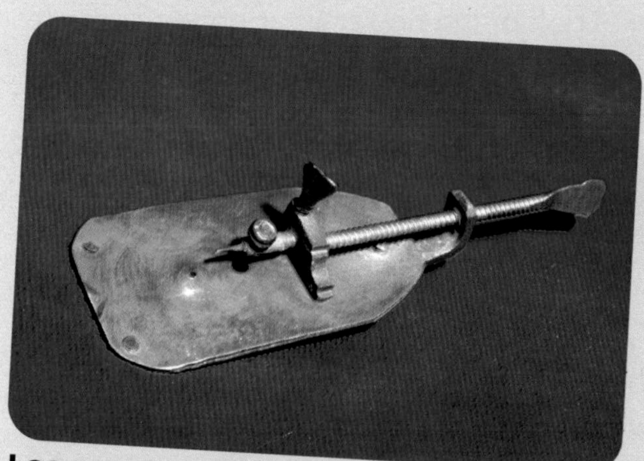

Leeuwenhoek's small microscope

will pollinate for nectar

- Use a microscope to see cells, and use stains to make them show more clearly
- Cells have a cell membrane, cytoplasm and a nucleus
- Plant cells also have cell walls, chloroplasts and a large vacuole
- Plant and animal cells are made by cell division
- Bacteria reproduce by binary fission

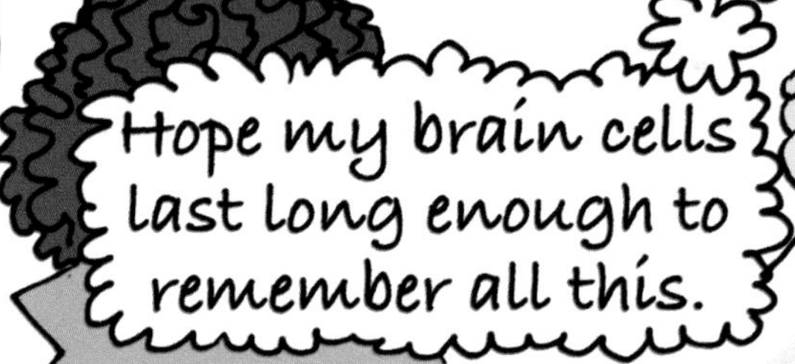

Hope my brain cells last long enough to remember all this.

DANGER! AVOID THESE COMMON ERRORS

In mathematics *multiplying* and *dividing* mean almost the opposite of each other. But when we talk about making more cells, or more bacteria, they mean the same thing! When cells reproduce, we use *dividing* because they appear to split into two. Yet they are *multiplying* at the same time because there are more cells at the end of the process.

Seeds are spread by the wind or carried by animals. This is called dispersal. Pollination deals with how pollen is carried from one plant to another.

Key words

divide
observation
pollination

UNIT REVIEW

REVIEW QUESTIONS

Understanding and applying concepts

1 Copy and complete the sentences, using words from the lists below.

The lenses of a ... magnify the image. To see cells, a ..., which is a thin slice of material, is cut and placed on a glass slide. It is covered with a ....

cover slip section microscope

Inside a cell you can see the ..., which is a rounded structure that controls the cell's activity. The cell is filled with .... Plant cells have a more regular shape because they have a ....

cytoplasm nucleus cell wall

A tissue is a group of specialised ... that carry out a particular job. New cells are made by .... A cell divides to make two new cells.

cells cell division

2 Draw a plant cell. On your diagram label the **nucleus, cytoplasm, cell wall** and **vacuole.**

3 Copy the outline of the human body. Put the letter of each organ in the correct place on the outline. At the side of the outline write what each organ does.

A heart
B lungs
C brain
D liver
E kidney
F uterus (womb)

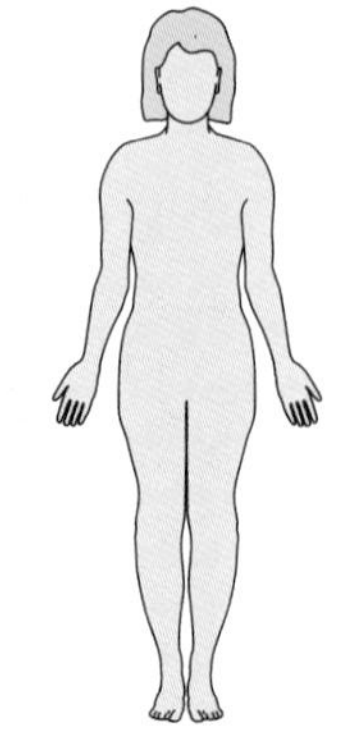

4 Organs can carry out their jobs because of the specialised cells they have. Match up the organs, specialised cells and the jobs they do.

organs: **root lungs anther leaf**
cells: **epithelium cell in the windpipe palisade leaf cell root hair cell pollen cell**
jobs: **fertilising an ovule making food removing dust absorbing water**

Thinking skills

5 Make a Venn diagram, like the one below, of the words and phrases. Group them into 'Animal only' (A), 'Plant only' (P), and 'Animals and plants' (A&P).

cell membrane cell wall chloroplast cytoplasm nucleus ovule palisade cell red blood cell sperm vacuole pollen root hair cell tissue organ egg

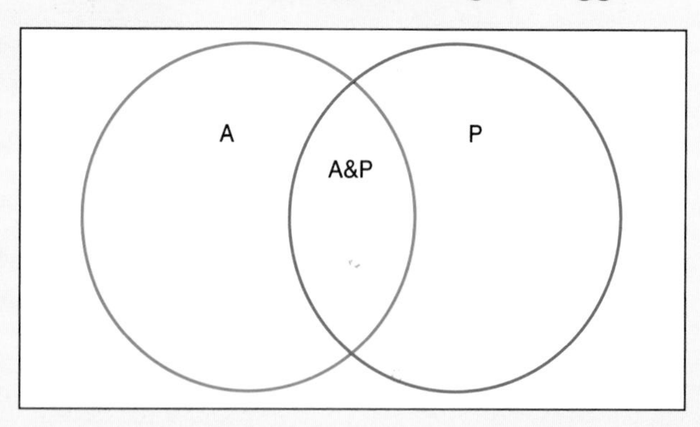

Ways with words

6 Physicists, chemists and biologists use the word 'nucleus' in different ways. Write a sentence using 'nucleus' as a biologist studying cells would use it.

SAT-STYLE QUESTIONS

1 The diagram shows a section through a wallflower.

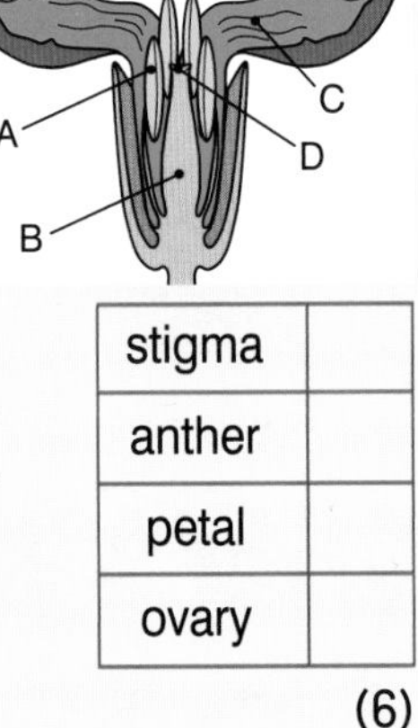

a The names of four parts of the flower are given in the table. Write the letter of each part on the drawing by its name.

stigma	
anther	
petal	
ovary	

b Which part produces pollen?

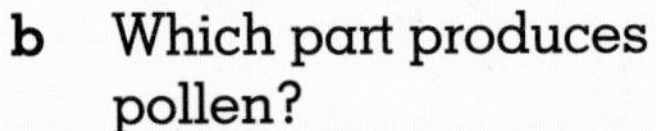

c Which part contains ovules? (6)

2 The names of five organs are given in the table. Choose the correct function for each organ from the list below and write it by the name.

digests food   holds developing baby
controls behaviour   absorbs oxygen
pumps blood

heart	
stomach	
lung	
uterus	
brain	

(5)

3 The diagram shows a cell from a leaf.

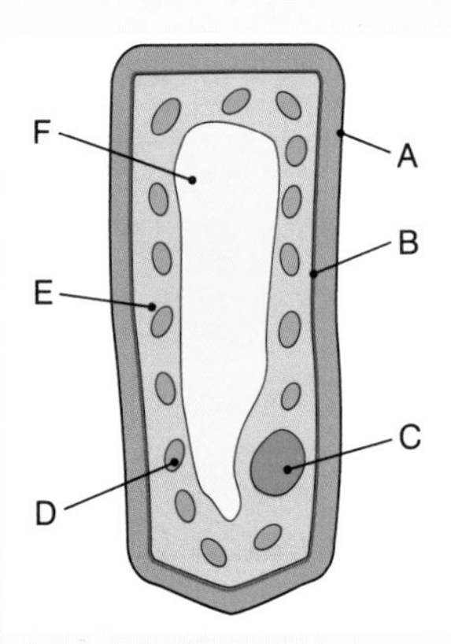

a Match the parts of the cell with a letter on the diagram. You will have one part left over.

cell wall   cytoplasm   nucleus
vacuole   chloroplast (5)

b Which parts are also found in animal cells? (2)

4 The diagram shows cell A from the lining of the windpipe. Its job is to sweep fluid carrying dust particles up and out of the lungs.

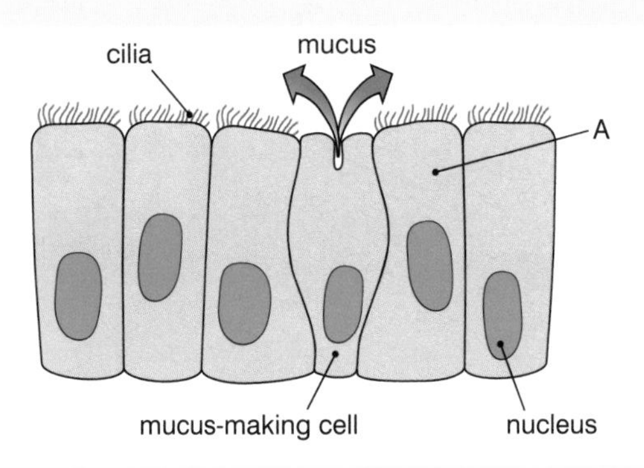

a What is the job of the nucleus?

b What word describes a group of similar cells that work together?

c How is the cell specialised for its job? (3)

5 Josh and Arun had been finding out about bees. They discovered that bees see blue, violet and ultra-violet light best. They wondered if bees were more likely to visit blue flowers.

They made model flowers out of filter paper. They coloured them with food dye and dropped sugar solution on the centre. They made some blue and some pink flowers.

Arun and Josh placed the model flowers near a flowerbed in the school grounds. They recorded how many bees visited the flowers.

Flower colour	blue	pink
Number of visits by bees	14	10

a Arun made a prediction. Which of these statements is a prediction?

A The flowers were made out of filter paper.
B Bees collect pollen from flowers.
C Bees will visit more blue flowers than pink. (1)

b Josh and Arun wanted to make this a 'fair test'. What do they need to think about? (2)

c Which colour flowers did the bees visit most frequently? (1)

d Do the data they collected support Arun's prediction? (1)

e Can you think of another explanation for their results? (1)

Key words

Unscramble these:
garon
elvous
lonple
suites
coal vue

7B

Reproduction

Coming soon

What's it all about?

Your life changes many times. You start school, you leave school, you get a job, you move house . . . . Your body changes too. Each new stage of life brings different activities, different ways of doing things, and different worries and concerns.

Somewhere between 10 and 13 years old you will begin the changes of **adolescence**. Your body grows and your feelings change. You learn to cope with new ideas.

By the end of this unit you will know more about how your body develops and how your feelings change during adolescence.

What do you remember?

You already know about:

- human and animal life cycles
- how plants reproduce
- cells

1 Sort these life stages into the right order.

teenager baby pensioner
parent child adult

2 When we grow we make new cells. Where do they come from?

a cell-making gland in our skeleton
from eggs by parent cells dividing
they sprout under your skin

3 Work in a small group. Take turns describing the life cycle of a pet to the rest of your group.

Ideas about growing up

QUESTIONS

The Scientifica crew are about to find out more about adolescence. What can you tell them?

- How does your body change as you pass from child to adult?
- How do your emotions change?
- Was Mike right? Do periods last for one day?
- What does a baby need to grow healthily?

7B1 Growing up

Learn about

- puberty
- human growth

Research shows that parents have a harder time during their child's **adolescence** than the child does.

Changes

You have grown up a lot in the last few years. You can mix with new people, and you are more independent. Your body has grown too. All these differences are part of **puberty**.

Growing

Boys and girls grow in the same way until they are about 12 years old, although girls start their growth spurts slightly before boys. After that, boys become taller and heavier than girls.

Changing

Puberty is the time of the last growth spurt. It starts sometime between 10 and 14 years old.

Your brain makes a **hormone** that starts it all off. A hormone is a chemical that controls the activity of other parts of the body.

Your skeleton and muscles grow so that you become taller. Inside your body your reproductive organs develop.

The growing reproductive organs make other hormones. These cause the changes we see in our body and in our interests, moods and emotions. The changes take until the late teens to complete.

We grow more fine hair all over the body, and coarser hair grows in our armpits and between our legs. Hairs also start to grow on a boy's face. We also begin to smell more strongly because our skin glands start to work.

Q1 What can you do to stay 'fresh as a daisy'?

brain hormone action, ovary, ovary hormone, wider hips; enlarged larynx, larger muscles, testis hormone, testis

The female sex organs

The male sex organs

Girls grow breasts and their hips widen. Boys' shoulders broaden and their voices deepen because their larynx (voice box) grows larger.

A boy's penis and testicles grow and start to produce sperm. Girls begin to have their periods.

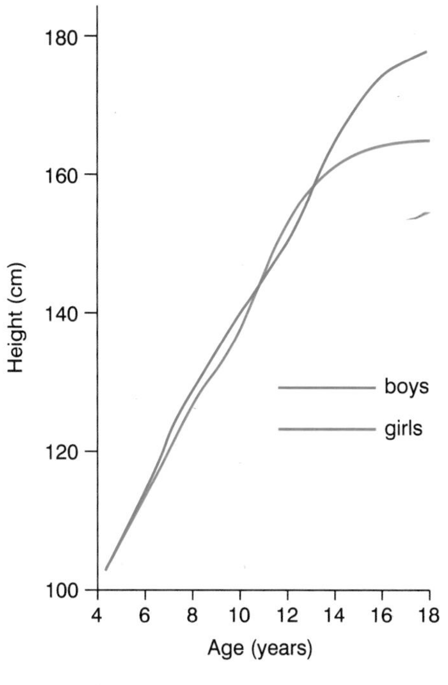

Typical growth curves for boys and girls

Have you grown?

- Measure the height of everyone in your class. Think about how to measure every person in the same way.
  - **a)** How are you going to set out your data?
  - **b)** What would be the best way to present your findings?
- Take the measurements again at the end of the year.
  - **c)** How much has everyone grown?
- Reese Cycle has a height chart on her wall. Her height was marked on it every 3 months. Use the data in the table to make a growth chart for Reese.
  - **d)** Does her growth pattern follow the usual pattern?
  - **e)** At what ages did she grow fastest?

Age	Height
6	107
$6\frac{1}{2}$	110
7	113
$7\frac{1}{2}$	117
8	120
$8\frac{1}{2}$	124
9	127
$9\frac{1}{2}$	129
10	132
$10\frac{1}{2}$	135
11	140
$11\frac{1}{2}$	144
12	149
13	156
14	160

SUMMARY QUESTION

1 ☆ Make three lists headed 'Boys', 'Girls' and 'Both sexes'. List each change under the appropriate heading.

sex organs get bigger **hair grows on face and chest**
breasts get larger **hair grows in armpits and between legs**
ovaries start to make hormones **voice gets deeper**
periods start

Key words

adolescence
growth
hormone
puberty

LEARN ABOUT

- the reproductive organs
- how sperm and eggs meet
- fertilisation

New individuals

Two special cells are needed to make a new life. They are a **sperm** cell, made by the father, and an **egg** cell, made by the mother.

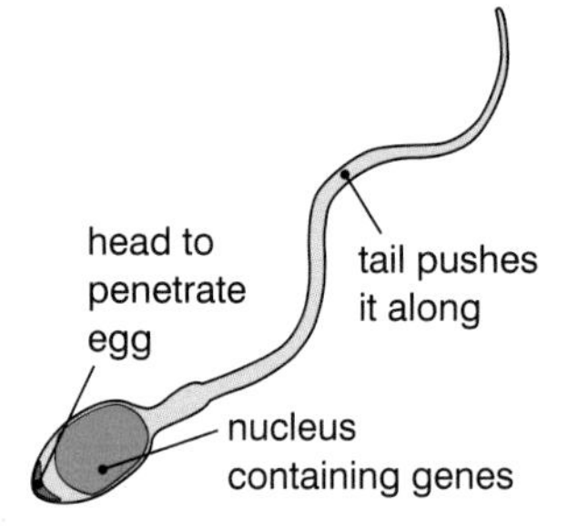

A sperm

Testes and sperm

Men make sperm in two **testes**. They are inside the **scrotum**, by the **penis**. Testes start to make sperm during puberty.

Sperm are very small cells. They are specialised for taking copies of a man's genes to an egg. Sperm cells have very little cytoplasm. Once they are released they survive for just a few days.

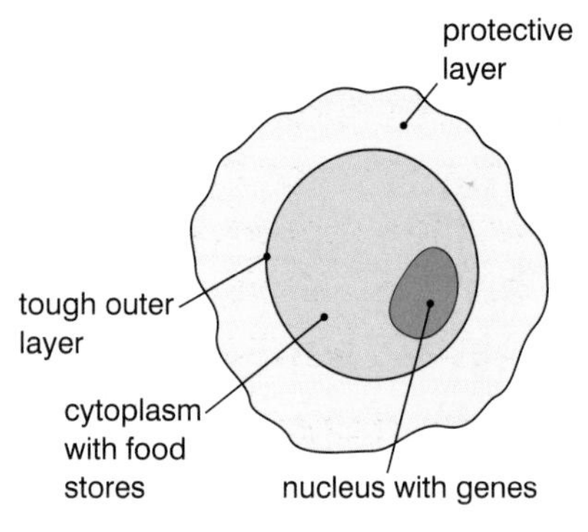

An egg

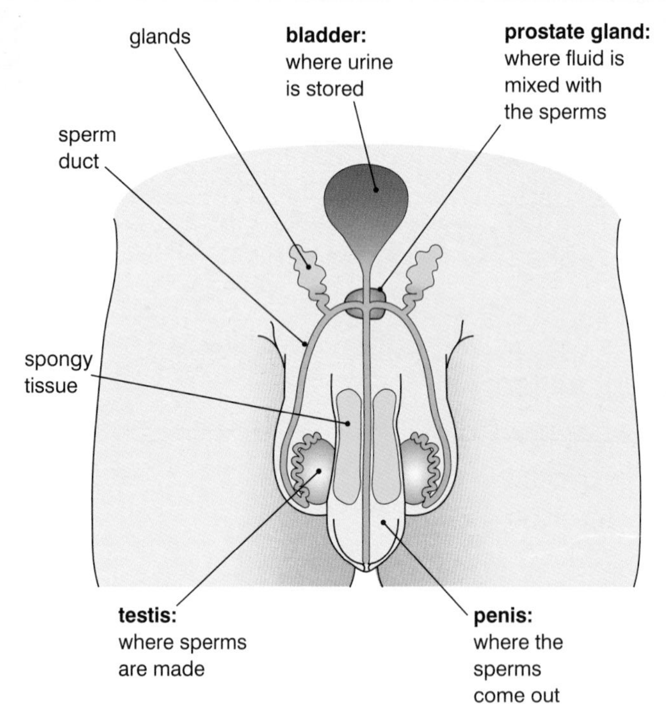

The male reproductive system

The tube in the oviduct is as narrow as a human hair.

Ovaries and eggs

Eggs are made by two **ovaries** in the woman's abdomen. Each month one ovary releases an egg. The egg travels along the **oviduct** towards the **uterus**. Egg cells are much larger than sperm cells and carry some food stores.

The female reproductive system

oviduct: where one egg passes to the uterus each month

uterus: where a fertilised egg grows into a baby

ovary: where eggs develop

sperm swim up towards the oviduct where they could meet an ovum

cervix

bladder: where urine is stored

sperm swim to the uterus from the vagina

vagina: where the baby comes out

Sperm meets egg

During sexual intercourse a man releases sperm inside the woman's vagina. His penis is usually soft, but it becomes swollen and erect so that it can be put into the vagina.

Sperm travel from the testes along the sperm duct. When the man ejaculates, he releases millions of sperm into the vagina near the entrance to the uterus.

Sperm swim through the uterus and into the oviduct to meet an egg. Hundreds of sperm reach the egg, but only one sperm fertilises it. The nucleus of this sperm cell passes into the egg cell. This is called **fertilisation**. The fertilised egg cell now has the genes it inherited from both parents.

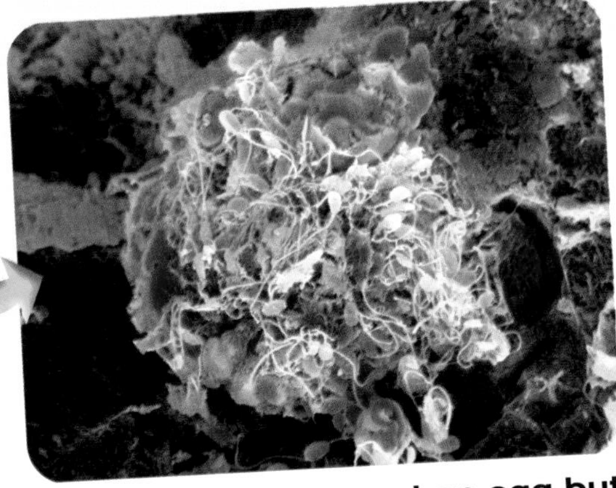

Many sperm surround an egg but only one fertilises it

Drawing paths

- Label the reproductive organs on a diagram of the human reproductive systems. Write each organ's function by its label.
- On a diagram of the female reproductive system, use blue arrows to show the path that the egg takes before it is fertilised. Use red arrows to show the path taken by sperm. Mark the place where fertilisation takes place with an X.

SUMMARY QUESTIONS

1 ☆ What is fertilisation?

2 ☆ What part of the cell carries genes?

3 ☆☆ Some couples find it hard to conceive a baby. Some men make too few sperm. Some women may have their oviducts blocked. Use what you have learned to explain how these can make it hard to conceive a baby.

Key words

ovary
oviduct
penis
sperm
testis (pl. testes)
uterus

7B3
Menstrual cycle

Menstrual cycle

LEARN ABOUT
- how eggs are produced

What is a period?

A girl's periods are part of her **menstrual cycle**. Each cycle lasts about 28 days, but it can be shorter or longer.

The first half of the cycle

The ovary contains thousands of undeveloped eggs. At the start of each cycle one egg starts to grow. The egg develops inside a small pocket of cells.

At the same time the **uterus** gets ready to receive a fertilised egg. The uterus is made of muscle. It has a lining that will help nourish a developing baby. In the first half of each cycle the uterus lining grows and thickens.

LINK UP TO
CITIZENSHIP

Sex and relationships can cause problems for older teenagers.

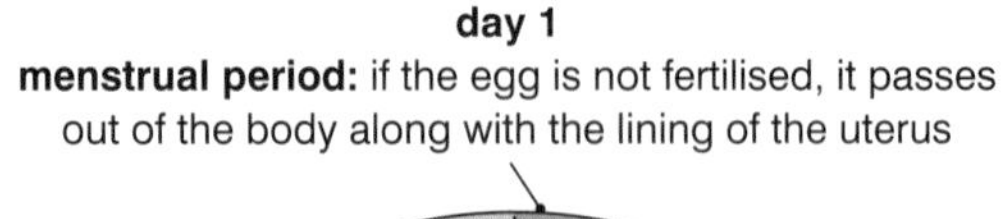

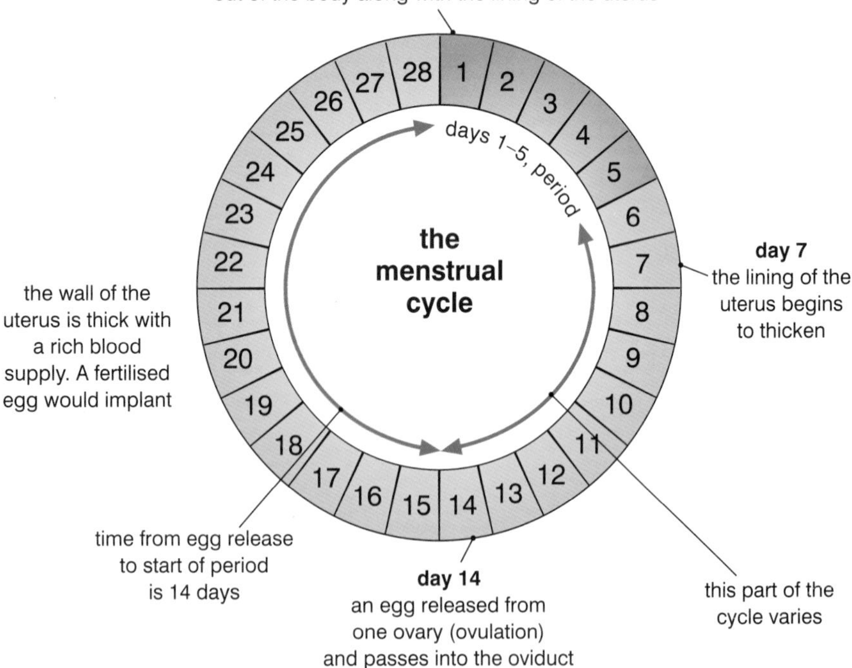
day 1
menstrual period: if the egg is not fertilised, it passes out of the body along with the lining of the uterus
the menstrual cycle
days 1–5, period
day 7
the lining of the uterus begins to thicken
the wall of the uterus is thick with a rich blood supply. A fertilised egg would implant
time from egg release to start of period is 14 days
day 14
an egg released from one ovary (ovulation) and passes into the oviduct
this part of the cycle varies
1 2 3 4 5 6 7 8 9 10 11 12 13 14 15 16 17 18 19 20 21 22 23 24 25 26 27 28

AMAZING SCIENCE!

The first human baby conceived outside the body – Louise Joy Brown – was born on 25 July 1978 in Lancashire.

Ovulation

When the egg is ready, it is released from the ovary. This is called **ovulation**. The egg enters the oviduct and travels down the tube. It lives for about a day after it has been released, unless it is fertilised.

The second half of the cycle

- *If the egg is fertilised* by a sperm in the oviduct, it will start to develop. When it reaches the uterus, the fertilised egg lodges in the lining of the uterus. The lining gets thicker. The cycle is switched off until after the baby is born.
- *If the egg is NOT fertilised* by a sperm, it dies. The thick lining in the uterus starts to break up. The lining is lost with blood through the vagina. This is called 'having a period', or **menstruation**.

The changes near the end of the cycle can make women feel irritable and uncomfortable.

By the time the period begins a new egg has begun to grow in the ovary. So menstruation also marks the first day of the next cycle.

Twins

Occasionally two eggs are started at the same time. If each of them is fertilised two babies begin to develop. These will grow into **non-identical twins**. These twins can be the same sex, or boy and girl. Having twins can run in the family. If you are a twin – watch out!

Monthly cycles

- Look at the calendar. It shows some of the dates in Amy Stevenson's menstrual cycle.

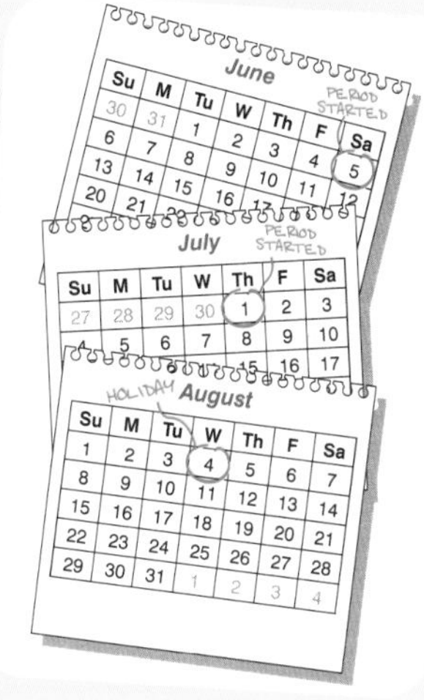

a) If her cycles are regular, when will she expect her next period to start?

b) On which days could she conceive a baby, if there were sperm in her oviducts?

- She is going on holiday on 4 August for 2 weeks.

c) Will she need to be prepared for having a period while she is away?

SUMMARY QUESTIONS

1 ☆ What is:
a) menstruation **b)** ovulation?

2 ☆☆ How might a woman realise that she is pregnant?

3 ☆☆ Why do girls need more iron for new blood cells in their diets than boys?

Look at the table showing the amount of iron in some foods. Choose two foods that would be good for a girl having heavy periods.

Food	Iron (mg per 100 g of food)
chocolate	1.6
cheese	0.4
baked beans	1.4
rice	0.5
spinach	3.0
beef	3.0

Key words

menstruation
ovulation
non-identical twins
uterus

7B4 Pregnancy

LEARN ABOUT
- how a baby develops
- the placenta

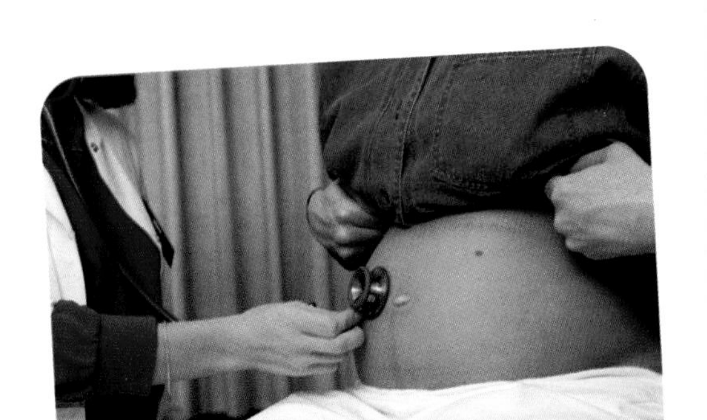

How does a baby develop?

After an egg cell has been fertilised it travels down the oviduct. It divides to become a small cluster of cells. When the cluster of cells arrives at the uterus, it attaches to the lining. It grows into a **foetus**. The foetus is not called a baby until it is born.

The foetus is protected by a membrane bag, called the **amnion**. The amnion is filled with **amniotic fluid** which supports and cushions the growing foetus.

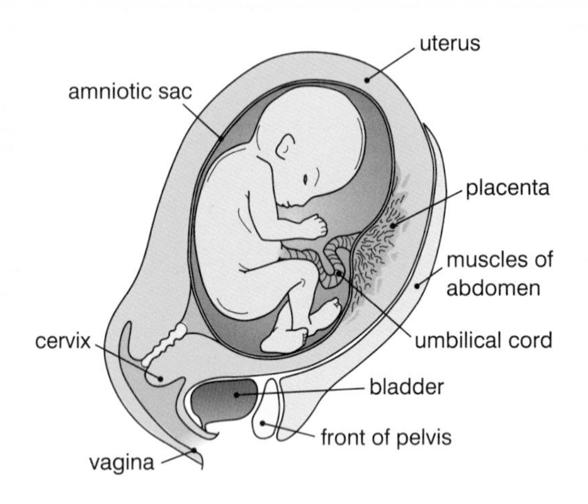

The time a foetus takes to grow from fertilisation to birth is called the **gestation period**. Human babies are born after 38 to 40 weeks of pregnancy.

The placenta

A foetus needs oxygen, water and food materials to grow. It obtains them from its mother through the **placenta**.

The mother's blood flows to the placenta, carrying all these useful materials. The foetus' blood travels to the placenta through the **umbilical cord**. Useful materials pass from the mother's blood into the blood of the foetus. Wastes pass from the blood of the foetus into the mother's blood.

Asian elephants have the longest pregnancy. It lasts for 650 days, which is nearly 21 months.

Q1 Where has the oxygen in the mother's blood come from?

The placenta keeps the foetus safe from harmful substances and microbes in the mother's blood. They cannot pass through it into the blood of the foetus.

...Choo
Bless you!

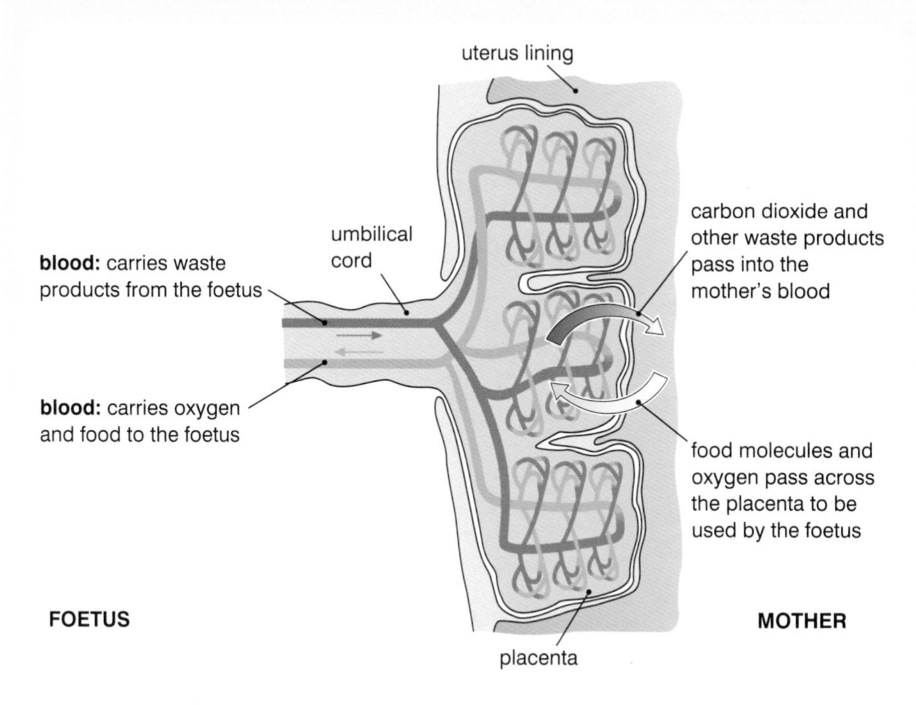
uterus lining
umbilical cord
blood: carries waste products from the foetus
blood: carries oxygen and food to the foetus
carbon dioxide and other waste products pass into the mother's blood
food molecules and oxygen pass across the placenta to be used by the foetus
FOETUS
MOTHER
placenta

The placenta

Identical twins

Occasionally, the ball of cells formed early in the pregnancy splits into two clusters of cells. Each group grows into a separate baby. The babies are **identical twins** because they have copies of the same genes in their cells.

Making a model

- Make a model of a developing foetus in the uterus using household materials.

SUMMARY QUESTIONS

1 ☆ Explain how identical and non-identical twins (see p.27) come about.

2 ☆☆ Describe how the placenta supplies and protects the foetus.

Key words

amnion
amniotic fluid
identical twins
gestation period
placenta
umbilical cord
foetus

LEARN ABOUT
- birth
- how a baby obtains its food

What happens at birth?

By the 38th week of pregnancy, a foetus has grown enough to be born. The important organs are working well enough for the baby to survive independently. The baby has a layer of fat under the skin which keeps it warm. Fat is also an energy supply once the baby is born.

The foetus turns so that it is head down. The uterus is held closed by a ring of muscle, called the **cervix**. During **birth**, the cervix opens to allow the baby to pass into the vagina. The baby is pushed out by very powerful contractions of the uterus.

The umbilical cord is cut after the baby has been delivered. The placenta is pushed out after the baby.

AMAZING SCIENCE!

Uterus contractions are one of the most powerful human muscle actions.

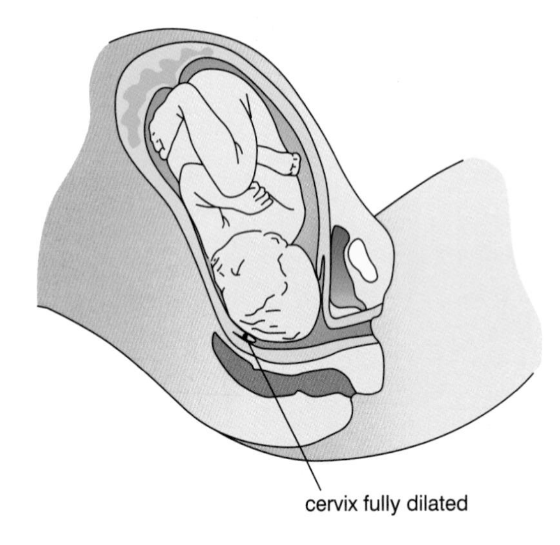

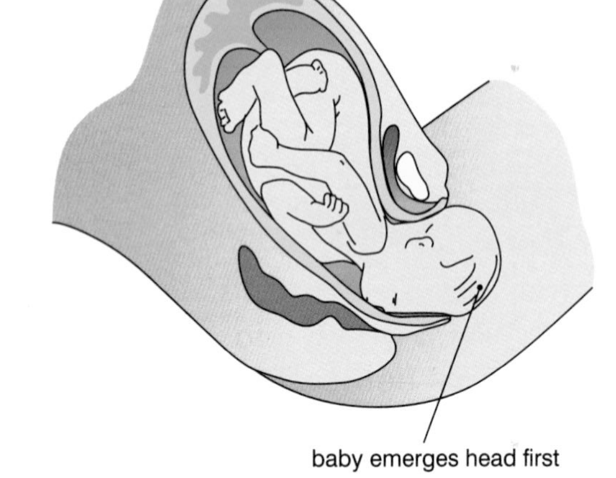

A baby being born

Baby care

The newborn baby needs milk. Milk is made in its mother's breasts by **mammary glands**.

Human milk has exactly the right balance of food materials for the baby. It also carries **antibodies**, which protect the baby from infectious microbes. The baby's digestive system will not be ready for other foods for several months.

Breast is best

Doctors and other health advisors say that 'breast is best'.

- Find out why breast-feeding is better for a baby.

Development

New babies find their food source by turning their heads when their cheeks are touched. Babies grow rapidly in the first few months, learning to control their limbs. At the same time, their sociability and thinking skills develop.

Within weeks they recognise the person who usually looks after them. They interact with people by crying, babbling, reaching and smiling.

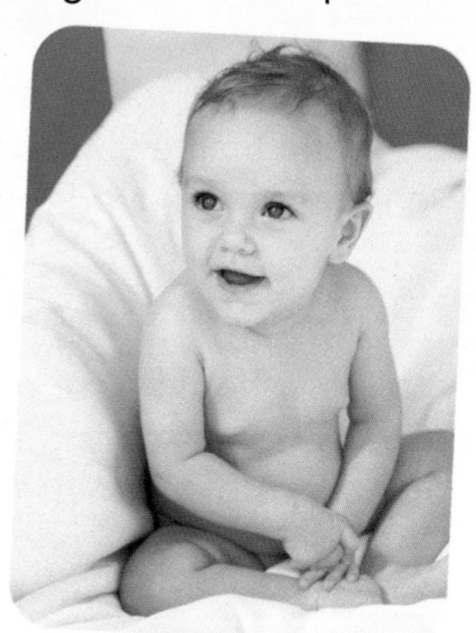

Babies enjoy being with people, being shown books and toys, and seeing the world. These activities help their development in childhood.

Special-care babies

Premature babies are born too soon. They need special care to survive because many of their organs are not fully grown. They are nursed in incubators which can give them extra oxygen.

They cannot breathe well because their lungs are stiff and need help. Their digestive system may not be ready for bottle milk. Many are fed with a solution of nutrients through a tube.

The incubator is full of warm air because premature babies cannot keep warm. They have very thin skin and no body fat.

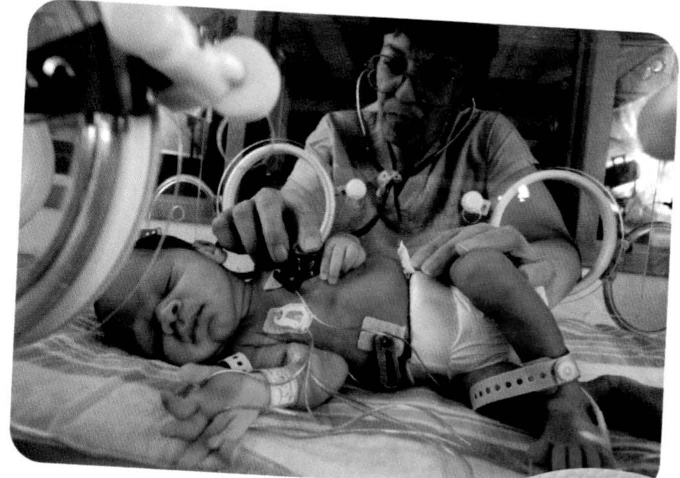

Premature babies have sensors to monitor the heart and how much oxygen is in the blood

SUMMARY QUESTIONS

1 ☆ How do newborn babies obtain the nourishment they need?

2 ☆☆ Give one advantage of breast-feeding.

Key words

antibodies
birth
cervix
mammary glands
premature babies

7B6 Healthy pregnancy

LEARN ABOUT
- things that can harm a foetus

What's the problem?

The developing foetus is very sensitive to harmful factors during the first few weeks. At this time a mother may not even realise she is pregnant!

Vitamins

A woman should have a good diet *before* she conceives a baby, as well as during the pregnancy. In this way she has good stores of vitamins and minerals. The baby may suffer if she does not have enough of some vitamins.

Q1 What should she eat to get more vitamins?

Microbes

Most microbes cannot pass through the placenta to the foetus. Unfortunately the **rubella** virus can pass through it. Rubella virus is also called German measles. The virus damages nerve cells and the foetus' heart. Children are vaccinated against rubella with the **MMR** (measles-mumps-rubella) vaccine. Now few women become infected.

Smoking

Cigarette smoke contains poisonous chemicals, such as **carbon monoxide** and **nicotine**. These affect the foetus and the course of the pregnancy.

The problems include:
- the foetus has less oxygen, and may be smaller at birth.
- the baby is more likely to be premature. Premature babies are more likely to have health problems.
- a bigger risk of complications in pregnancy.
- a bigger risk of the foetus dying during pregnancy.

Some of the effects of smoking on a baby's developments last for several years

Table 1: Effects of smoking on birth weight. Babies of mothers who smoke weigh less than babies of mothers who don't smoke

Country where the research was done	Brazil	USA Study 1	USA Study 2	USA Study 3	Norway Study 1	Norway Study 2	Japan
How much more babies of non-smokers weighed compared with babies of smokers (in grams)	142	18	377	348	182	197	96

Drugs and alcohol

Doctors take special care with medicines for pregnant women. Some drugs harm the developing baby. Pregnant women are also advised to cut down the amount of alcohol they drink because it can also have harmful effects.

Other sorts of illegal drugs can also pass through the placenta and may harm the baby.

Drinking more than four alcoholic drinks a week can affect the foetus

Designing posters

- Design a poster to persuade pregnant women to give up smoking or to reduce the amount of alcohol they drink during pregnancy.

Some of the harmful effects of drugs and smoking are discussed in Unit 9B.

SUMMARY QUESTION

1 ☆☆ Look at Table 1. It shows the results of several studies on the effect of smoking on a baby's birth weight. Each shows that babies of smokers weigh less than those of non-smokers.

a) How much more, on average, do non-smokers' babies weigh?

b) What are the problems of having a very small baby?

c) Describe the ways that smoking can affect the developing foetus.

Key words

MMR
nicotine
carbon monoxide
rubella

How animals reproduce

LEARN ABOUT

- how other animals reproduce

Do all animals reproduce in the same way?

Fish and frogs produce hundreds of young. Cheetahs and whales have just a few. There are many different patterns of reproduction in the animal kingdom. Each has advantages and disadvantages.

Mammals

A **mammal** egg has a good chance of becoming a baby. The baby also has a good chance of becoming an adult, so mammals do not need to have thousands of young. Mammals such as cheetahs keep their egg cells inside their body. Eggs are likely to be fertilised because sperm is placed inside too. This is **internal fertilisation**.

Young mammals develop inside their mother's body, where they have a constant supply of nutrients. They are warm and protected from predators.

After birth young mammals are cared for by their mother until they can live on their own. They are protected from dangers and learn about finding food. They have a good start in life.

Pigs are good mothers and protect their young from predators

Investigating reproduction

- Choose three different animals to investigate: for example, a trout, a crocodile and a locust.
- Find out how they reproduce and make a presentation comparing their patterns of reproduction.

External fertilisation

When fish or frogs mate, the males and females get very close to each other. They release large numbers of eggs and sperm into the water at the same time. The sperm swim to the eggs to fertilise them. This is called **external fertilisation**.

Sticklebacks make a nest which improves their eggs' chance of surviving

The eggs are usually left to develop by themselves. Only a few of the thousands of eggs released by a female each year ever survive to become adults. Many die, or are eaten by predators.

Q1 Can you explain how birds reproduce?

Crocodiles are very good mothers and carry their young down to the water in their mouths after hatching.

Insects

Insects use internal fertilisation and lay eggs. They do not grow in the same way as we do. They have a hard, outside skeleton that stops them from getting larger.

Insects pass through growth stages. At the end of each growth stage, the hard skeleton splits and the young insect climbs out. It has a new, soft skeleton and some new features.

It expands itself as much as it can until its new skeleton has hardened. This leaves it 'room to grow' inside the new skeleton. Once it is mature, it does not grow any more.

The ladybird larva does not get wings and shiny red wing cases until it is an adult

SUMMARY QUESTION

1 ☆☆ Draw two overlapping circles to make a Venn diagram. Label one 'Lays eggs' and label the other 'Internal fertilisation'.

Put the animals in the list below into the diagram. Put the animals which have internal fertilisation but lay eggs with shells in the place where the circles overlap.

chicken salmon cobra leopard spider ladybird robin deer goldfish cat human turtle frog crocodile

Key words

external fertilisation
internal fertilisation
mammal

How scientists collect and use information

It has been hard to prove that cigarette smoke or not enough vitamins can harm babies. This is because scientists can't test their ideas by doing experiments on pregnant women. Babies could be harmed by experiments. This would be **unethical**. They have to use other methods.

Scientists collect information about babies with a health problem, such as spina bifida, and their families. Spina bifida is a condition that affects the nerves to the lower part of the body. The scientists look for **links** and **correlations** in the data. For example, spina bifida is more common in babies born at certain times of the year.

The researchers develop several **ideas** about the cause of the problem. For example, spina bifida might be caused by pregnant women eating fewer fresh vegetables at certain times of the year. But, because it is colder at certain times of the year, it might also be caused by pregnant women not getting as much fresh air and sunshine.

Ideas are tested as much as possible, sometimes using animals. Much more detailed information about affected babies and their family's lives and work is collected.

The data is analysed very carefully in order to eliminate other factors that could be contributing to the problem. Once the data supports one explanation more than the others, then remedies can be tried.

In fact spina bifida is linked to a shortage of a vitamin called **folic acid**. This is found in leafy green vegetables and some other foods.

- growth follows a pattern from birth until adulthood
- egg cells are produced by ovaries
- sperm cells are produced by testes.
- fertilisation occurs when a sperm nucleus enters an egg.
- the menstrual cycle is a monthly activity that produces eggs.
- egg production stops during pregnancy.
- puberty and the menstrual cycle are controlled by hormones.

I'm too young to read this.

Word Game

Link the letters together by making a word, the first has been done for you.

```
u n b o l t
t         e
e         s
r         t
u         e
s         s
```

DANGER! AVOID THESE COMMON ERRORS

Sperm are cells that can move independently. Eggs are also cells. Testes and ovaries are organs that produce these special cells.

Key words

correlation
folic acid
unethical

UNIT REVIEW

REVIEW QUESTIONS

Understanding and applying concepts

1 Copy and complete the sentences, using words from the list below.

Animals reproduce using special sex cells called . . . and . . . . These cells are produced in special organs. The male . . . are found in the scrotum, near the penis. The female sex cells are made by a pair of . . . located in the abdomen. Each month a single egg is released. Fertilisation takes place in the . . . . The fertilised egg makes its way to the uterus where it . . . in the wall. If it is not fertilised, the lining of the . . . is discharged.

testes oviduct eggs lodges uterus ovaries sperm

2 Describe the changes that boys and girls go through at puberty.

3 Look at the cell below.

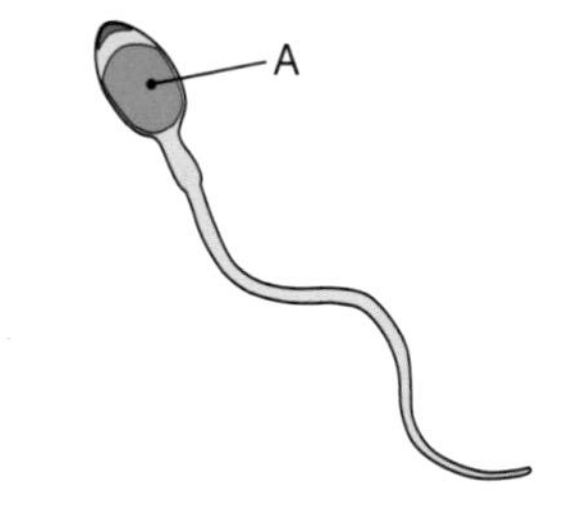

a What type of cell is this?
b What is its job?
c Give one feature that helps it do its job.
d What is part A?

4 Describe how a foetus obtains oxygen and nutrients from its mother. How does it get rid of carbon dioxide?

5 What is the difference between **internal fertilisation** and **external fertilisation**. Give one advantage and one disadvantage of each method.

6 Bob and Alice have decided it is time to start a family. Alice has a regular menstrual cycle of 28 days. She started her last period on 6 October. On what days in October is she most likely to conceive?

Thinking skills

7 Work in pairs to decide which is the odd one out in the groups below. Write down your reason for your decision.

a	**testes**	**ovary**	**sperm**
b	**infant**	**teenager**	**growth**
c	**menstruation**	**uterus**	**ovulation**
d	**moustache**	**breasts**	**wide hips**

SAT-STYLE QUESTIONS

1 This diagram shows part of the female reproductive system:

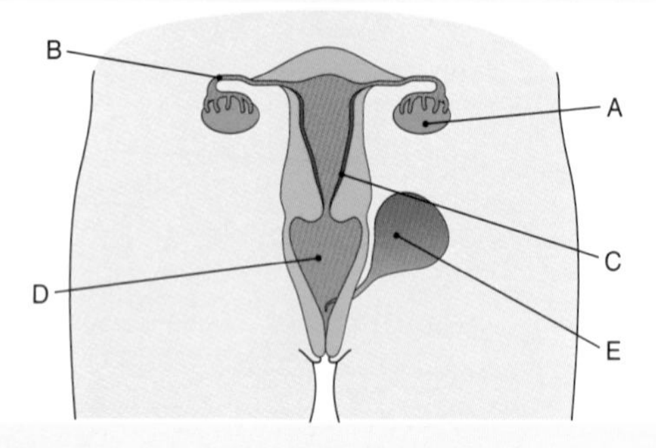

a Where does fertilisation take place?
b Where are the sperm deposited?
c Where is the egg cell released?
d Where does the fertilised egg start to grow into a foetus? (4)
e What happens at fertilisation? (1)

2 Look at the diagram of a developing foetus.

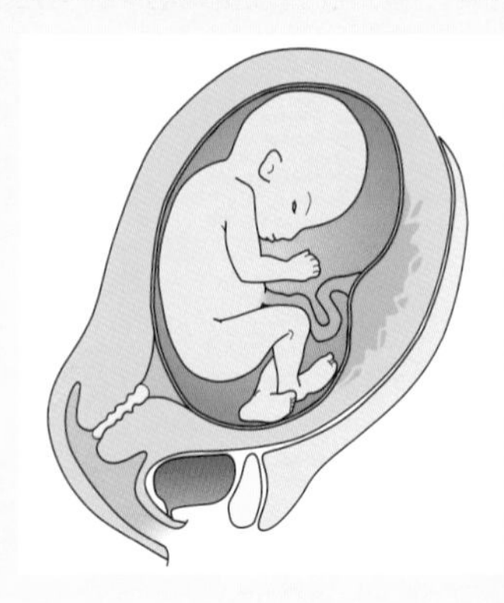

a Where does the foetus get the food and oxygen that it needs?
b How is the baby pushed out during birth? (2)
c Describe how a baby obtains its food after it is born. (2)

3 These are some substances that can pass through the placenta.

oxygen carbon dioxide digested food poisons from cigarette smoke water

a Name one substance that passes from the mother's blood to the baby's blood.
b Name one substance that passes from the baby's blood to the mother's blood.
c Name one substance that is harmful that passes from the mother's blood to the baby's blood. (3)

4 This question is about puberty and the menstrual cycle.
a During puberty girls' bodies change. Describe two ways in which they change. (2)
b Choose the correct word from the list below to complete each sentence.

ovulation uterus menstruation one oviduct four

i) _________ is when a girl has a period
ii) _________ is when an egg is released from the ovary
iii) The menstrual cycle takes about _________ weeks to complete
iv) Each cycle _________ egg is/eggs are released
v) At the end of each cycle the _________ lining is discharged from the vagina (5)

5 Lizzie kept pet rats. She bought two young female rats when they were 6 weeks old. She measured Millie and Coco from nose to tail tip every Sunday.

Week	Millie's length (cm)	Coco's length (cm)
1	24.0	22.5
2	25.5	24.0
3	27.5	25.5
4	29.0	26.5
5	—	—
6	32.5	29.5
7	34.5	31.0
8	36.0	32.0

a Which rat was the longest? (1)
b Use the information in the table to plot a graph.
- Put the number of weeks on the horizontal axis – this is the **independent variable**.
- Put the length of rat on the vertical axis – this is the **dependent variable**.
- Label each axis and write the units on each. (5)

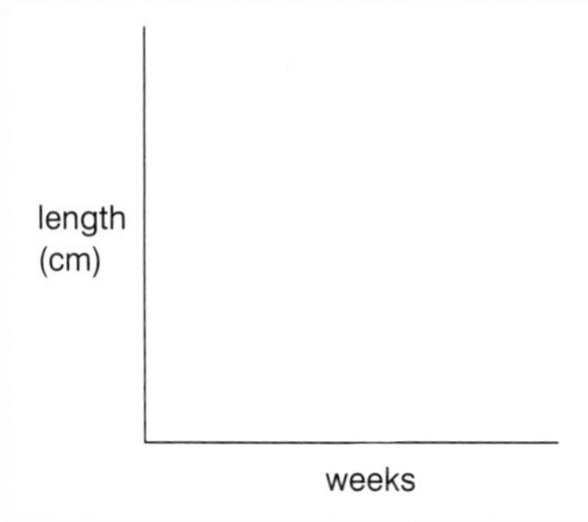

c Lizzie had to spend one weekend with her Grandma and she could not measure her rats.
d Use your graph to predict the length of each rat in Week 5.
e Give one reason why Lizzie's measurements might not be accurate.

Key words

Unscramble these:
pet canal
pry tube
vain tuloo
tse tse
suture

7C

Environment and feeding relationships

COMING SOON

What's it all about?

What's your favourite food? Chips? Cola? Feeding is a feature of life. Food gives us energy. All plants and animals need energy.

Plants use light to make their food. Some animals eat plants. Some animals eat other animals. Plants, the animals that eat them and predators, are linked together in **food chains**.

In this unit you will find out more about how animals and plants depend on each other. You will also learn about how animals and plants are adapted to survive in their habitat.

What do you remember?

You already know that:

- the animals and plants living in one environment are different from those living in another
- animals and plants depend on each other to live
- some animals feed on other animals
- some animals feed on plants

1 Name an animal which is a:
carnivore herbivore predator prey

2 Match the animal to the place where it lives.
animal: frog shark camel polar bear squirrel
place: desert fresh water woodland sea Arctic

3 Arrange these animals and plants into a food chain. Put the plant at the start of the chain. Link them with arrows.
bird lettuce slug cat

What's eating what?

QUESTIONS

- Can you tell the name of the feeding pattern in the song?
- Can you explain why predators need sharp teeth and claws?
- Can you think how being slippery could help a fish survive?
- Can you explain how living underground helps desert animals survive?

Learn about

- physical factors
- how organisms are adapted to their environment

Table 1: The main factors affecting animals and plants
how hot or how cold it is
how much the temperature changes each day
how much water there is in the soil
how much drinking water there is
how much it rains or floods
if there is enough oxygen in the surroundings
how much light there is
the chemistry of the soil

What's the problem?

Polar bears have thick fur. This helps them to keep in body heat in the cold Arctic climate. The low temperature is a **physical factor** that affects anything living in Arctic regions. There are many physical factors that can affect plants and animals. You can see these factors in Table 1.

Life in water

When you float in the swimming pool you feel light. Water helps to support your mass. Animals and plants that live in water do not need strong skeletons like land organisms. Water supports them.

Streamlining helps animals to move more easily through the water. This is because streamlining reduces water resistance and friction.

The barracuda's fins help it control movement in three directions. What do you think they are?

Fish absorb oxygen dissolved in water through their gills. Seals take in air before they go under the surface. They return to the surface to breathe again.

Which of these sea creatures breathes air?

shark turtle jellyfish prawn

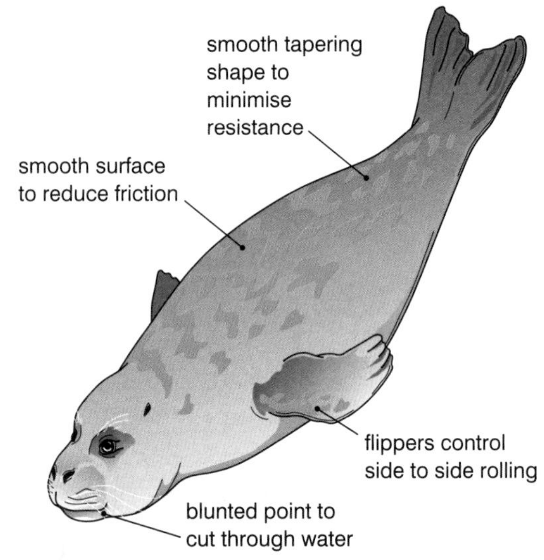

How a seal is adapted to move through the water

These barracuda are predators. They are streamlined enough to chase their prey at 40 to 50 km/h.

Life in the air

It is easy to move fast through air because it does not offer much **resistance**. The smooth surface of a bird's feathers reduces friction.

The fish eagle on p. 52 is a fast streamlined predator.

Animals that live in air must be very light. Birds and bats have a small body that weighs just a few grams. Birds have hollow bones that are very light. Their feathers are also light. Feathers are good insulators. This helps to keep a bird's tiny body warm.

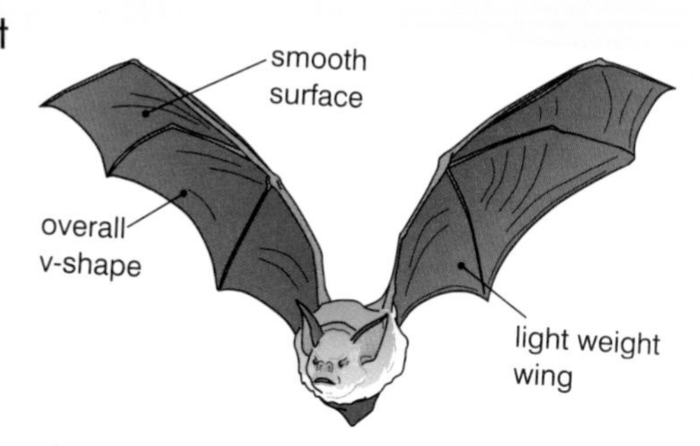

Bats' bodies are designed for lightness and fast flying

Life in the desert

Heat from the sun dries up water. Desert animals, such as the oryx in the Arabian Desert, can go without drinking for a long time. Desert plants have a thick, waxy or hairy outer layer that helps to stop water evaporating from their leaves. Animals stay in burrows where it is cooler during the day.

A cactus has a tough, thick outer layer. What do you think the spines are for?

ICT CHALLENGE

Find out about animals and plants that live in hot deserts. Make a poster of the features that help them survive.

Heat energy is transferred by conduction, convection and radiation.

An Arctic fox's fur keeps it as warm as wearing eight sets of clothes at once.

Smooth movers

- Use modelling materials to make model animal shapes. Drop them down a long cylinder of water. What is the most streamlined shape?

SUMMARY QUESTIONS

1 ☆ Look at the barracuda in the photo. Which features can you see that help it catch its prey?

2 ☆☆ Water animals that feed on plants in the water live near the surface. Predators and animals that feed on dead remains live lower down in the water? Why do you think this is?

Key words

environment
physical factors
streamlined
resistance

LEARN ABOUT

- measuring physical factors
- how light and water affect where a plant can grow

What is it like around your school?

We don't have to travel to the Arctic to find a tough environment. The plants and animals living around your school face plenty of problems in their habitat. What is their habitat like?

- There are sheltered corners with less wind near the buildings.
- It is warm close to buildings.
- There are ponds and puddles for animals to drink from. Birds drink and bathe in puddles on a flat roof.
- School buildings cast shade.
- Buildings stop rain reaching the ground.
- Rain water cannot pass through tarmac and concrete.
- The sports pitch has plenty of light (and rain!), but the grass is trampled. **Trampling** breaks plant leaves.

CHEERS!

Q1 List four physical factors mentioned in the paragraph you have just read.

Q2 Identify the physical factors that are likely to affect the plants below.

CHALLENGE

Use data-logging equipment to monitor physical factors around your school.

What plants grow near your school?

Do you have gardens at your school, or a wildlife area? Even if you do not have a nature area, there will be wild plants growing in the grounds. Animals and plants live in lawns, flowerbeds, trees and hedges. They also live in neglected areas.

Grass and small seedlings sprout at the bottom of a wall, in gutters and on the roof. These have grown from seeds that arrived from the surrounding area.

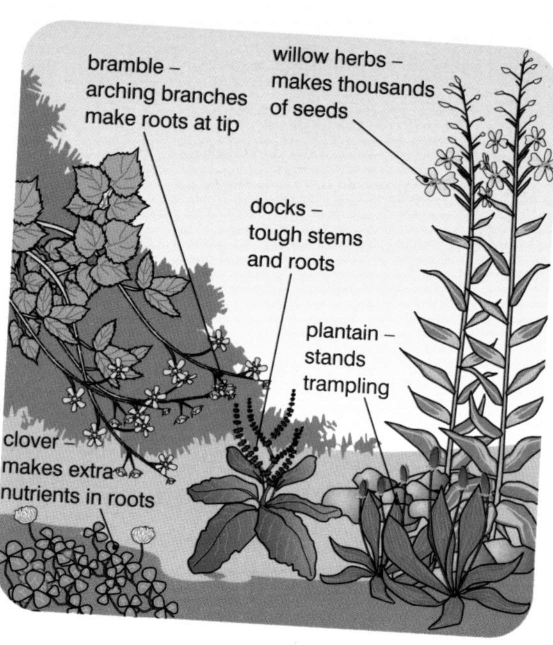

These plants have features that help them grow in tough places

Q3 How do seeds get into your school grounds? Choose the INCORRECT answer. Are they:

a) blown by the wind?
b) carried in mud on shoes and car tyres?
c) carried by insects?
d) dropped in bird droppings?

You are most likely to find plants that make huge numbers of seeds. Thousands of seeds arrive every year but only a few land in places where they can grow.

Physical factors

- Find and identify the plants growing in a sunny place and in a shady place in your school grounds.
  a) Are there any plants growing in one place but not the other?
- Measure and record the physical factors in the sunny and in the shady place. The important factors are:
  a) how much light there is
  b) how moist the soil is
  c) how **moist** the air is – this is called the **humidity**
  d) how warm the air and the ground are
  e) whether the soil is acid or alkaline
  f) how windy it is
- Compare the measurements you have made in the two places.
  g) Are there any differences?
  h) Why do you think there is a difference in the plants you found in the two places?

SUMMARY QUESTIONS

1 ☆ What instruments would you use to measure:
a) rainfall? **b)** wind speed? **c)** air temperature?

2 ☆ Name two **physical factors** that affect plants.

3 ☆☆ Why does **shade** affect a plant's growth?

Key words

humidity
moist
physical factors
shade
Trampling

7C3 Daily changes

Learn about
- how animals are adapted to daily changes

The wood mouse does not go out in bright moonlight in case a predator sees it

Day and night

During the day there is plenty of light. Animals can use their eyesight to find food. Sparrows and crows may look for crumbs near the school litterbins. Predators such as kestrels look for prey such as mice on the ground. It is warmer in the daytime too. Cold blooded animals like frogs and toads are active.

Things change when it gets dark. Plants need light to make food. Dandelions growing around your school cannot make food at night. It is also cooler and the dandelion's flowers close.

Some animals are active only at night. They are **nocturnal**. Cats and owls are predators that hunt at night. They have large eyes to see in the low light levels. Nocturnal predators use other senses too for detecting prey. A cat's large ears detect rustling in vegetation, and they have a good sense of smell.

Tides

There is another sort of change at the coast. The tide moves in and out twice each day. At high tide the shore is covered in sea water. At low tide it is open to the air.

Animals and plants living on the shore risk drying out at low tide. The slippery coating on seaweed helps to prevent it drying out. Animals hide in the sand or under rocks to keep moist.

Limpets and mussels cluster on the sheltered side of rocks away from pounding waves

At high tide, waves buffet the animals on rocks and seaweed. Limpets clamp themselves tightly to the rocks to stop themselves being washed away.

CHALLENGE

Set up a video-cam to monitor animal life at night in one of the quieter but well-lit areas of the grounds.

Which animals live near your school?

There are several ways you can find out.

- Set a **pitfall trap** to catch small animals living close to the soil.
- Look for insects on the leaves and stems of plants.
- Use a hand lens or binocular microscope to look for small animals living among leaf litter on the soil.
- Watch and record visiting birds. Look for signs of larger animals, such as tracks and gnawed food.
- Find out about the way of life of the animals you have identified.
- Enter your data on a database.

Millipedes are adapted to live in leaf litter. They can have up to 200 pairs of legs.

SUMMARY QUESTIONS

1 ☆☆ A pitfall trap is used to catch beetles and other small creatures. They fall in as they walk over it.

A pitfall trap was set up in a wild part of the school grounds one evening and emptied the next morning. Pip and Reese identified and counted the animals in it.

They noticed that there were more carnivores than herbivores in their trap.

Can you think of a reason for this?

2 ☆☆ Look at the owl. What features can you see which show that the owl is a predator and hunts at night?

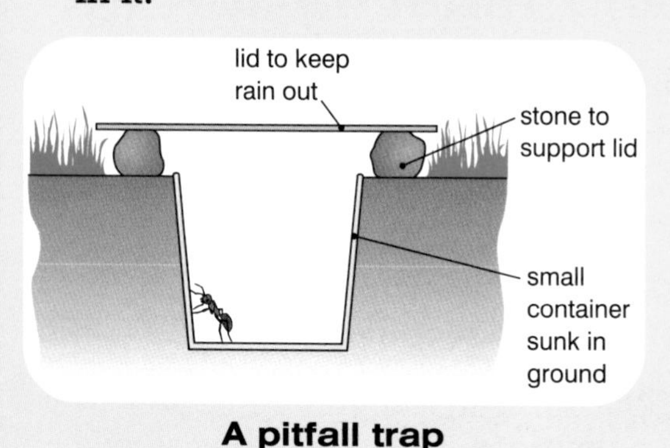

A pitfall trap

An owl

Key words

nocturnal
pitfall trap
tides

LEARN ABOUT

- how animals cope with harsh weather
- migration and hibernation

LINK UP TO PHYSICS

Our seasons are caused by the earth moving round the sun (Unit 7L.)

How the environment changes during the year

Summer is warm and light. Plants grow quickly. This means that there is plenty of food for animals that eat plants. Animals have their young in summer when there is plenty of food.

Q1 What do we call animals that eat plants?

Foxes grow thicker fur in the autumn

In winter it is colder. Sometimes water freezes in the soil, and in ponds and streams. It is windier and wetter than the summer.

The days are short, and there is not much light. Plants cannot make much food in these conditions. Their growth slows down so there is less food for animals.

Birds and mammals have to search for enough food to keep up their body temperature. Their fur or feathers insulate them.

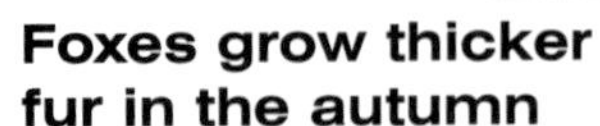

Bats feed on insects. They hibernate during the winter when there are few insects to eat.

Migration and hibernation

Swallows flying south in winter to a warmer environment. This is called **migration**. Animals in cold regions migrate further south, where the weather is milder and they can find food.

Frogs and lizards are not active in winter because their body temperature is too low.

Hedgehogs store fat in their body in the autumn. Fat is an energy store and helps to **insulate** the body. They slow down their bodily activity as they **hibernate** in a safe place.

Q2 Think of an animal that migrates to the UK from further north in the winter.

Q3 Think of one that migrates from the UK to further south in the winter.

Dormancy

In the autumn the leaves of many plants die but the parts underground stay alive. In its first year, a carrot plant stores food in its main root, the carrot. In the autumn, the leaves die and the carrot plant becomes **dormant**. It survives on stored food, but does not grow. When the ground warms up, the carrot grows new leaves.

Carrots, potatoes, parsnips and onions survive the winter underground

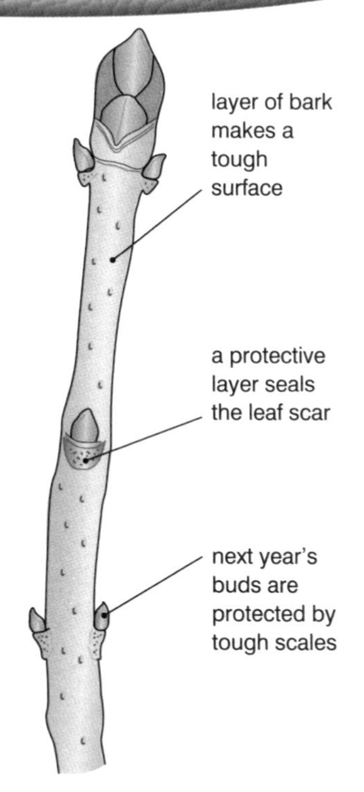
layer of bark makes a tough surface
a protective layer seals the leaf scar
next year's buds are protected by tough scales

Trees protect their delicate buds from winter frost

Investigating insulation

- Investigate how well wool and feathers insulate a flask of hot water.
- Animals have a layer of oil on their hair. This stops water wetting them. Find out what happens to insulating materials when they are wet.

Gruesome science

A hibernating grizzly bear loses up to half its body weight in pure fat during the winter months. No wonder it wakes up starving!

SUMMARY QUESTIONS

1 ☆ Stoats are mammals that hunt rabbits.
   a) How does its coat help it survive the winter?
   b) In Scotland it grows a white coat in winter, but in the Midlands it stays dark brown. What do you think is the advantage of this?

2 ☆☆ Hibernating animals may die after they have been woken by an unusually warm spell in the winter. Can you explain why?

Key words

dormant
hibernate
insulation
migrate

What eats what?

LEARN ABOUT
- why food chains start with plants
- food webs

Food chains

The animals and plants living in a wood are linked together because they depend on each other for food. Rabbits eat grass for food. Foxes eat rabbits. We can link the grass, rabbits and fox together in a **food chain**.

All of the animals and plants in a woodland can be linked into food chains. When we write out a food chain, we use arrows to show which way the food energy goes. You can see this in the diagram.

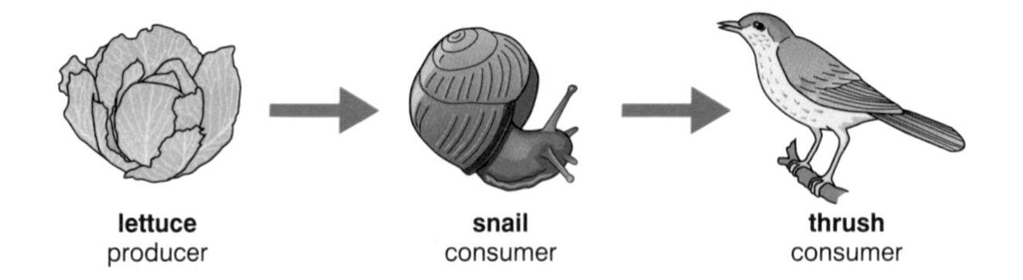

A food chain

All food chains begin with plants. Energy passes from the plants into the animal that eats them, and so on.

Making food chains

- Use the information you found about the animals and plants in your school grounds to link them together in food chains.
- Use any observations you made which could help to identify a food source. For example, you might have found a house fly in a spider's web.
- Try to explain any missing links.

Producers and consumers

Plants are called **producers** because they fix light energy in the foods they make.

Greenfly, rabbits, kangaroos and elephants are called **consumers** because they consume the energy in plants. Predators such as foxes, lions, spiders and eagles are also consumers. They consume the energy in their prey.

Consumers consume the energy in their prey

Food webs

Animals eat more than one sort of food. Each animal and plant can be a part of several different food chains. Slugs in your garden eat cabbages, lettuces and flowers as well as dead plants. Foxes eat slugs, worms, apples and mice as well as rabbits.

A food chain showing lettuce → slug → fox is just one of many routes by which energy can pass from one living thing to another. A **food web** shows more feeding links between animals and plants in a community. The diagram shows a food web in a woodland.

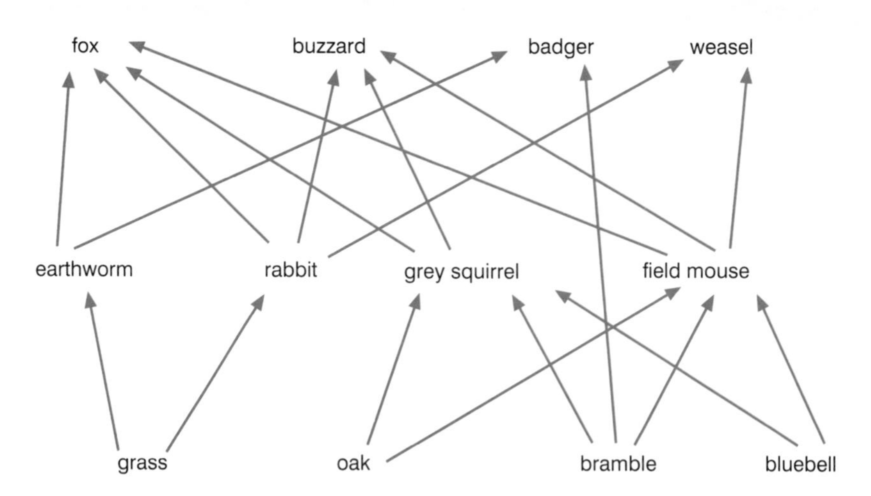

A food web in a woodland

Plants are all producers so they are all placed on the lowest level of the diagram. Animals that eat plants, called herbivores, are on the next level up. At the top are the predators.

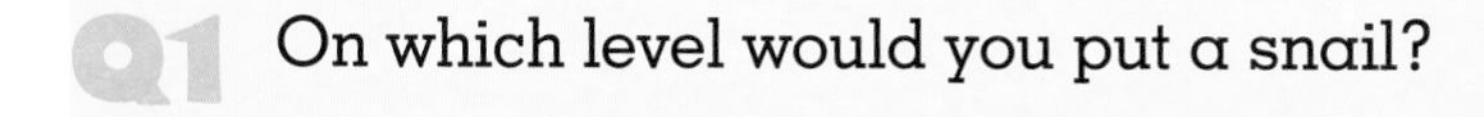

Q1 On which level would you put a snail?

AMAZING SCIENCE!

An oak tree can support a food web with as many as 200 different sorts of living things in it.

SUMMARY QUESTIONS

1 ☆ Look at the diagram of the food web above.
   a) Pick out three food chains from the food web.
   b) Identify a producer, a consumer, a herbivore and a carnivore in the food web.

2 ☆☆ Find two different ways in which energy can pass from a bluebell to a fox.

Key words
consumer
food chain
food web
producer

LEARN ABOUT

- how plants defend themselves
- features that help predators catch prey
- how prey animals avoid predators

A fish eagle catching a fish

Predators

Predators have special features to find, catch and kill their prey.

The fish eagle in the photograph feeds on fish. It uses its sharp, forward-facing eyes to spot fish. Its long legs reach under the surface for the fish. It hooks the slippery fish with large sharp curved **talons**. The eagle's powerful wings lift it clear of the water again. It has a curved beak to tear meat from the fish.

Q1 Predators may stalk, trap or ambush their prey. Think of a predator that uses each of these methods.

ICT CHALLENGE

Find out how a zebra's stripes, a tiger's stripes, a ladybird's bright colour, and a hedgehog's spines help them survive.

The prey

The fish that the fish eagle hunts are **camouflaged**. They are dark grey on the top. This makes them hard to see against the bottom of a lake. Underneath, the fish are silvery white. Predators at the bottom of the lake don't see them because they are camouflaged against the silvery surface.

Prey animals are good at detecting predators. Rabbits have large eyes on the sides of their head to see in all directions. Their large ears have sensitive hearing. Even a shadow passing overhead will alarm rabbits. If a predator is near, they run fast, using their powerful back legs.

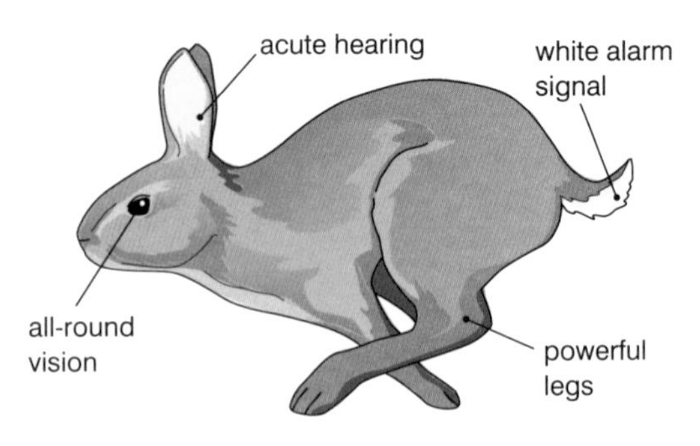

Rabbits are well-adapted to avoid predators

Plants

Being stung by a nettle is painful and you have learnt to avoid them. Nettles and other plants have fine hairs, prickles, chemicals, thorns or spines to stop animals eating their leaves.

Herbivores tug plants out of the ground when they feed. Dandelions and dock have long, tough roots to keep them anchored. Daisies and dandelions keep their growing point close to the ground, where it is less likely to be eaten.

Wild guinea pigs live in long grass. They detect predators by smell and acute hearing. If you have a guinea pig, it can hear and smell you long before it can see you!

The daisy's hairy leaves grow in a flat ring. They are less likely to be damaged.

A parrot's beak is strong and pointed to crack hard nuts

Testing the effectiveness of camouflage

- Make model caterpillars out of wool or modelling materials.
- Put them in a particular area of the grounds. Challenge another group to find them.
- Choose one predator and one prey animal. Find out about their specialisations.

SUMMARY QUESTIONS

1 ☆ Look at the parrot. How is it adapted to eat large seeds.

2 ☆☆ Make a list to compare the features of predators and prey. The first example has been done for you.

Predators	Prey
sharp eyes to spot prey at a distance	good all-round vision

Key words

camouflage
predator
prey
talon

7C7 Disturbing the web

LEARN ABOUT

- how changes in a food web affect the other animals
- how animals compete for food

Changes in a food web

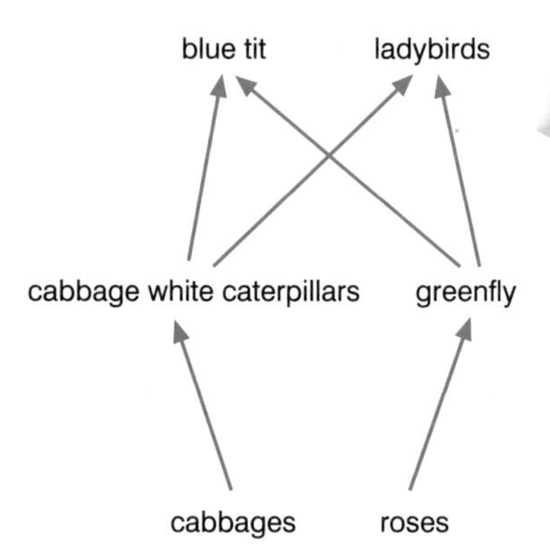

A food web in the garden

You know that animals eat several kinds of food. Hedgehogs eat worms, slugs, beetles and small frogs. If there aren't many slugs, they will eat more worms and beetles.

But when hedgehogs are eating more beetles, there are fewer beetles for other animals to eat. Shrews also eat beetles. If hedgehogs are eating more beetles, then shrews will not have enough food. The hedgehogs and shrews are **competing** for food. There may not be enough beetles for all of them to thrive.

Look at the food web in the diagram. If the number of cabbage white caterpillars decreases, the animals that feed on them will have to eat more of a different food.

Q1 What will the animals feeding on caterpillars eat instead?

Q2 Which two animals will be in competition?

Roads are a danger to breeding toads

Why do the numbers of animals change?

The numbers of animals can change for many reasons.

- When there is plenty of food, more animals can survive. When there isn't very much, the number goes down.
- Animals need safe places, like hedges, to breed and to escape from predators. When farmers dig up hedges there are fewer safe places.
- Animals and plants have a smaller area to find food and mates when new roads and houses are built.
- Animals and plants are sensitive to chemicals used on farms, in factories and in houses. They can damage animals and plants if they get into the environment.

Conservation

Some animals and plants need special living conditions. Otters need clean rivers with plenty of quiet places along the banks. Water pollution and flood-control measures reduced the number of suitable places and otters disappeared from many areas. Cleaning up rivers has helped otters to spread again.

Farmers drain marshes to improve the land for cattle or crops. Animals like the raft spider, which is only found in a few very marshy areas, have almost disappeared If we want to **conserve** rare animals and plants, we have to protect the places they live and the food chains that they depend on.

CHALLENGE

Choose an endangered animal and use the Internet to find out how they are protected.

Food chains

- Look at the food webs you made of the animals and plants in the school grounds. Work out what will happen if you alter the numbers of some of your organisms.

Finding a place for tigers away from people is getting harder

SUMMARY QUESTIONS

1 ☆ Look at the food web on p.54. What will happen if the number of caterpillars decreases.

2 What will happen in the food web shown:

a) ☆☆ if the farmer uses insecticide to kill the flies?

b) ☆☆☆ if the farmer decides to dig up the hedges to make three fields into one big field?

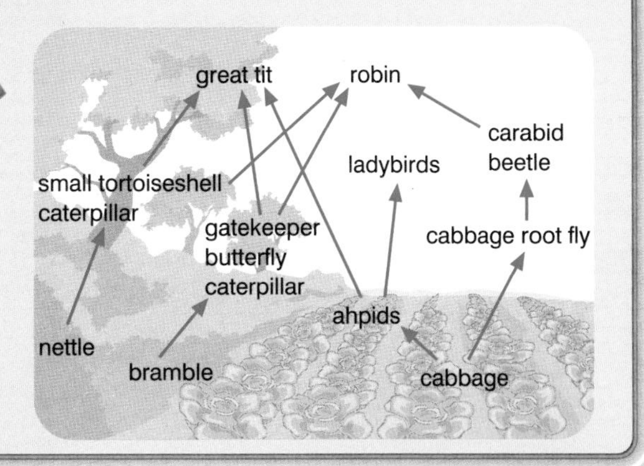

Key words

competing
conserve
food web

How do we know what eats what?

Ecologists investigate how plants and animals live together. Animals are often very shy, and so it is difficult to find out what they are eating. Ecologists study traces that animals leave in order to find out what is eating what.

Small animals usually live on the food they eat. Butterflies lay their eggs on the plants their caterpillars eat. So, when you see a caterpillar on a plant, it is usually eating it.

Ecologists can find out what animal ate a nut by looking at the shape of the teeth marks on the nut shell. Finding broken snail shells around a stone tells us that thrushes are about. Thrushes break open snails by dropping them on a stone.

Looking at droppings is a good way to find out what an animal has eaten.

Droppings full of fish bones tell us otters are about

Owls cough up pellets that contain bones and indigestible bits of the animals that they have eaten. We can work out what an owl eats by looking at these remains.

We can also look at the stomach contents of dead animals. But scientists working with sharks have to remember that sharks don't just swallow prey. They will even swallow plastic containers and lifebuoys as well as tin cans!

SCIENTIFIC PEOPLE

Park rangers manage areas of countryside on the edge of a large town or city. Sutton Park in the Midlands is a lowland heath with lakes and ponds full of wildfowl. Part of it is a **Site of Special Scientific Interest (SSSI)**. This kind of habitat has almost disappeared, so it has to be looked after. At the same time, hundreds of people use it every day for riding horses, walking dogs, cycling, orienteering and other activities.

Rangers try to balance the needs of people with the needs of wildlife. They look after the woods so that they do not take over the open heath, and keep the ponds and rivers clear. They monitor and encourage the wildlife. They also run play areas and a visitor centre where people can find out about the Park. Most park rangers have a background in estate or countryside management, biological sciences or ecology at university.

• the key physical factors in an environment are temperature, water, oxygen, and light
• organisms are adapted to survive within their environment
• animals are adapted to be active at night and for changing seasons
• animals are adapted for finding their food and avoiding being eaten
• food chains start with plants
• food chains are linked together into food webs
• factors which affect organisms in one part of a food web affect others too.

DANGER! AVOID THESE COMMON ERRORS

The arrows in a food chain show how food energy passes along the chain. Light energy passes into plants from the sun. Plants use it to make foods.

When an animal eats a plant, it takes in food energy. In this way, the energy passes from the Sun to animals.

Make sure your arrows point the right way. If the arrows point the wrong way you are saying that lettuces eat slugs, and greenfly eat robins! Your food chain should look like this:

sun —(light energy)→ plant —(energy stored in the plant)→ herbivore —(energy stored in the animal)→ carnivore

Key words

ecologist
Site of Special Scientific Interest (SSSI)

UNIT REVIEW

REVIEW QUESTIONS
Understanding and applying concepts

1 Copy and complete the sentences, using words from the lists below.

A . . . . . . shows all the feeding links between the animals and . . . in a community. Plants use light energy to make foods. They are . . . . Animals that eat the plants take in energy stored by the plants. Animals that eat plants are called . . . .

herbivores producers food web plants

Predators are animals that catch and . . . other animals. They have keen . . . to detect their prey. The animals they eat often have good . . . and a keen sense of smell to warn them when a predator is near. Predators move fast to catch prey. Sharks are . . . so they can move quickly. . . . hunters have large eyes so that they can see when there is not very much light.

nocturnal hearing streamlined senses kill

2 Can you think of a reason why:

- **a** a red deer grows a longer, shaggier coat in the winter?
- **b** a blackberry plant has thorns?
- **c** a hover fly has a black and yellow striped abdomen, like a wasp? (A hover fly is a harmless fly that feeds on nectar in flowers.) (3)

Thinking skills

3 Choose an animal and make a spider diagram of the ways in which it is specialised for its life. Think about:

- how it moves
- where it lives
- how it is copes with the seasons
- what it feeds on
- how it avoids being eaten

4 The map below shows a river. Scientists investigated the animals living at three places. These places are marked A, B and C on the map.

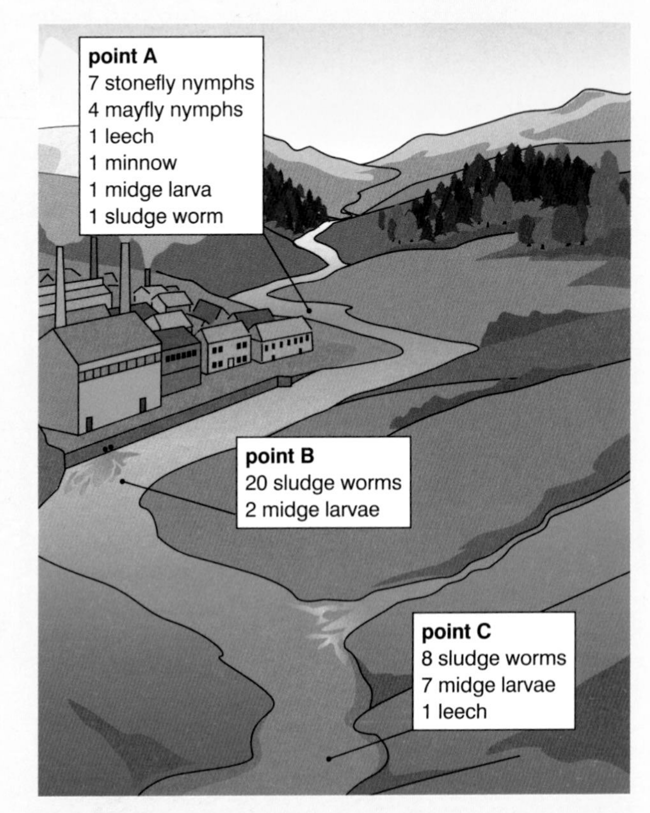

- **a** Which animals were found in all three places? (2)
- **b** Which animals were only found in the cleanest water? (3)
- **c** Which animals were best able to live in polluted water? (2)
- **d** Which substance, necessary for life, is likely to be scarce at place B? (1)

SAT-STYLE QUESTIONS

1 The drawings show four living things.

Robins eat caterpillars, caterpillars eat cabbages, sparrowhawks eat robins.

a write down the food chain which has sparrow hawks, cabbage, robins and caterpillars in it (3)

______ ⟶ ______ ⟶ ______ ⟶ ______

b Which living thing in this chain is a producer? (1)

c Give the name of a predator in this food chain, and its prey.

predator ______

prey ______ (2)

2 Sea turtles are reptiles. They spend their lives at sea, eating seaweed and jelly fish. Female sea turtles come ashore each year to lay their eggs.

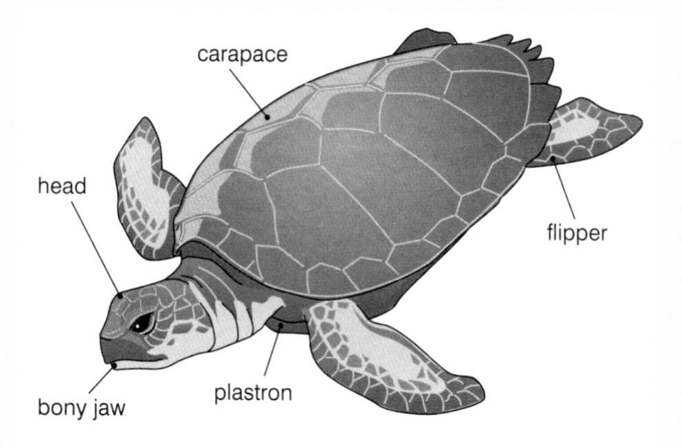

Match each fact about the sea turtle with the way it helps the turtle to survive

its beak is sharp and horny	This protects the soft body from predator bites
it's flippers are broad	This helps it to breathe the air above the sea surface
its nostrils are near top of its head	can bite through tough seaweed
The scales are fused together to make a hard carapace	this helps it to swim fast through the water

(4)

5 Reese and Pip were investigating the animals living in the area behind the Year 7 classrooms. They had found woodlice in a pile of bricks by the fence but there weren't any woodlice under a bush nearby. Reese thought that it was not dark enough for woodlice under the branches. They decide to investigate why there were woodlice in one place but not the other. They set up a choice chamber for woodlice as shown below.

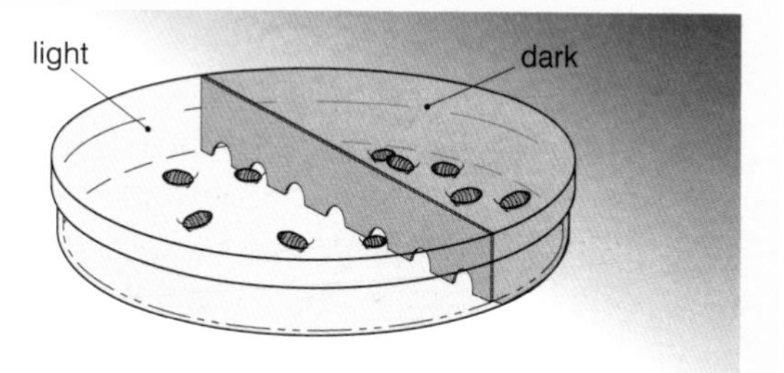

A choice chamber

They placed 10 woodlice in the centre of the chamber and left them for 5 minutes. After this time they counted how many woodlice were in each part of the chamber.

a write down the question the pupils were investigating (1)

b what would Reese expect to see? This is her prediction. (1)

They found that 7 woodlice were in the dark part and 3 were in the light part.

c draw a bar chart of the data Reese and Pip collected (3)

d do the data support Reese's prediction? (1)

Key words

Unscramble these:

mun score
bait hat
rim gate
to draper
drup core

7D Variation and classification

Coming soon

What's it all about?

You share your life with animals and plants. They look very different, but when we look carefully we can see that they share some features. For example, both your hamster and garden birds have a backbone – and so has your goldfish.

Scientists sort living things into groups using the features they share. You know about mammals already but do you know why whales and dogs are mammals but a shark is a fish?

You will find out about inheriting features and how farmers breed animals and plants.

What do you remember?

You already know:

- the names of parts of a plant
- how to use a key to identify an animal or plant

1 Match A, B, C and D with the correct label?

leaf
root
stem
flower

2 Use the key to find out what sort of animals these are.

1 Has wings Go to **2**
Doesn't seem to have wings Go to **3**

2 Has a large pair of jaws at the head end ... stag beetle
Has a pair of pincers at the rear end ... earwig

3 Has one pair of wings with legs longer than the body ... crane fly
Has two pairs of wings with a long, pointed spike at the rear ... horntail

Differences and similarities

QUESTIONS

The Scientifica crew are about to find out more about the differences between groups of animals and plants. What can you tell them?

- Has Pete guessed correctly?
- Mike didn't believe the puppy was possible because he thought the parents were too different. Is he correct?
- How would Benson explain how the baby looked like its parents
- How could Pip and Mike find out what the creepy-crawly is?

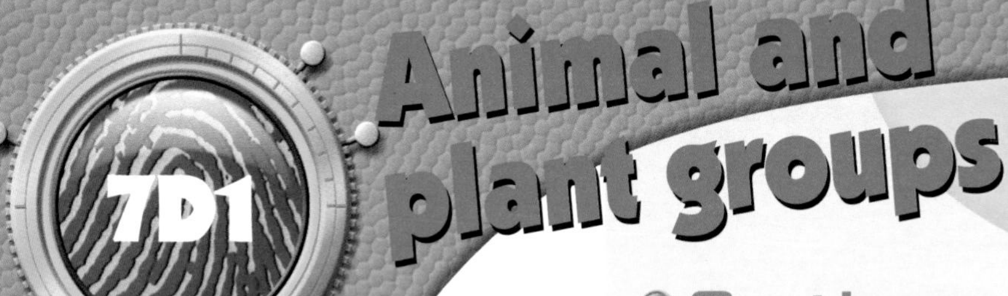

Animal and plant groups

Learn about
- sorting living things

Sorting

We have found millions of different kinds of living things as we have explored the planet – and we are still discovering new ones. Scientists have sorted these millions into groups of similar animals and plants.

There are many different ways of sorting animals and plants. A wildflower guide sorts plants to help a reader identify a flower more quickly. It might sort flowers into pink ones and blue ones. Or it might sort them into the months when they flower.

There are between 10 and 30 million species but no one knows for sure.

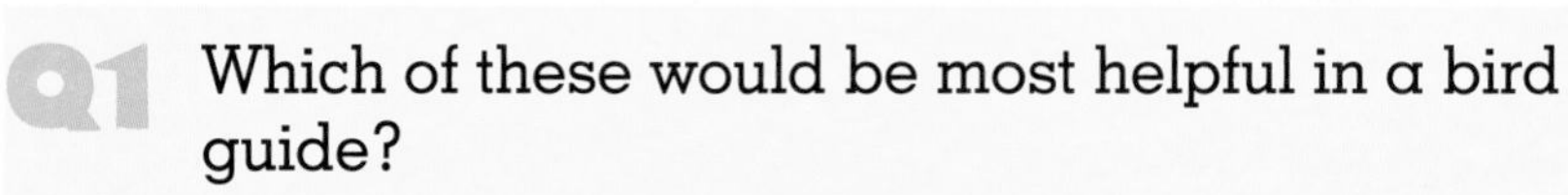

Q1 Which of these would be most helpful in a bird guide?

size of bird   where you saw it
its song or call   colour

Scientists sort animals and plants using information about their body structure and how they reproduce. All the members of a group share the same typical features.

Q2 The lion, tiger, lynx, jaguar, panther and domestic cat are all in the same group. What features do they share?

The groups

There are five **kingdoms** of living things, shown in the pictures. Each kingdom is divided into smaller groups. For example, animals are divided into those which have a backbone and those which don't (see Section 7D2).

The five kingdoms of living things

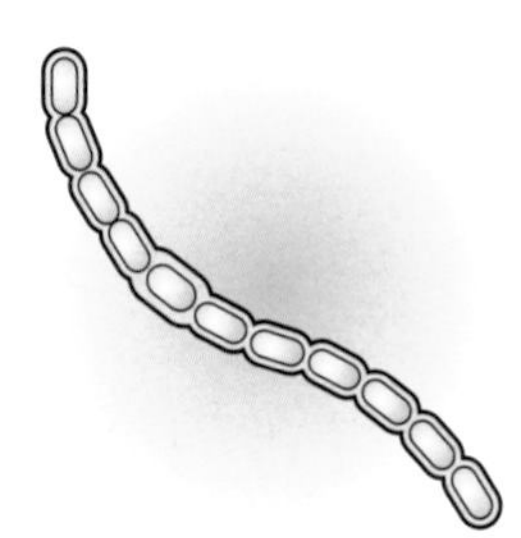

Monera – the bacteria and their relatives

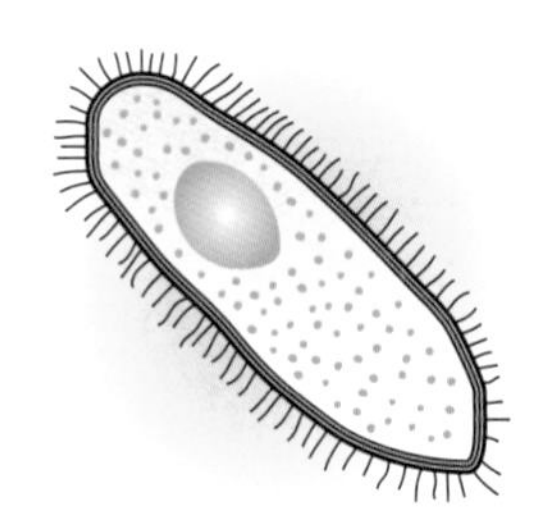

Protoctista – organisms with just one cell, such as Paramecium

Fungi – moulds, mushrooms and their relatives

Animals – organisms that move about and have a nervous system

Plants – organisms that make food by photosynthesis

Sorting and describing

- Sort the animals shown in the pictures you have been given into groups.
  a) Why did you sort them in this way?
  b) Were any of your animals hard to place?
  c) If so, why was that?
- Now sort them in a different way.
- Look carefully at the specimen you have been given. Use a hand lens or binocular microscope to see small features.
  d) Describe its features. Try to make your description objective.

Like tulip flowers, some things come in a variety of colours, so be careful using colour for sorting

What is a species?

Animals (or plants) that belong to the same **species** can freely breed with each other and produce young. They cannot breed with members of another species – even if they look very similar.

Each species has a unique scientific name. For example, the bluebottle fly is called *Calliphora vomitoria*. As a result, scientists all over the world discussing *Calliphora vomitoria* know exactly which fly they are talking about.

Classification

Classification is the science of sorting organisms into groups. Carolus Linnaeus (1707–1778) started the scientific classification system. The animals, or plants, in a group are similar. They all evolved from the same ancestors over millions of years.

SUMMARY QUESTIONS

1 ☆ What is a 'species'?

2 ☆ Give an example of one member of each of the five kingdoms.

3 ☆☆ Why do you think a poodle and a spaniel can breed together but a chicken and goose can't?

Key words

classification
fungi
kingdom
species

LEARN ABOUT

- sorting animals

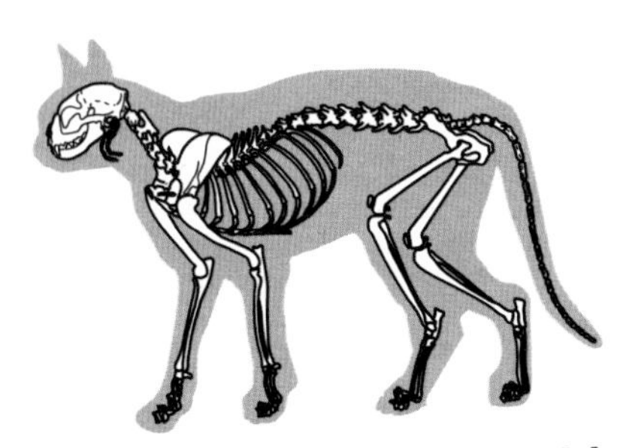

The cat is a vertebrate with a skeleton inside its body

Sorting animals

The animal kingdom contains animals as different as tiny soil mites and huge blue whales. About 5% of animal species are **vertebrates**. Vertebrates have:

- a backbone of bone or cartilage that protects the main part of the nervous system
- a body that is symmetrical from side to side
- a head and a tail.

There are five groups of vertebrates. These are the fish, the amphibians, the reptiles, the birds and the mammals.

The locust is an invertebrate that has a hard skeleton outside its body

The other 95% of animal species are **invertebrates**. Invertebrates do not have an internal skeleton, although some have a hard skeleton outside the body.

There are three times as many insect species as all the other groups put together.

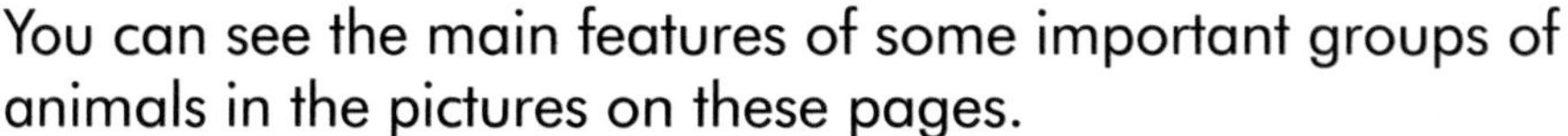

You can see the main features of some important groups of animals in the pictures on these pages.

Once we know some of an animal's features, we can work out which group it belongs to and what other features it is likely to have. For example, if it is hairy, four-legged and has warm blood, we can predict that it will probably give birth to live young and suckle them on milk.

Fish

- live in water
- have moist skin covered with scales
- breathe using gills
- move using fins
- lay eggs in water

Amphibians

- have moist skin without scales
- have four limbs but spend time in the water
- lay eggs and young develop in water

Reptiles

- have four limbs at an angle to the body
- have dry skin with scales
- lay eggs with a leathery shell

Birds

- fly using two wings
- have feathers
- lay eggs with a hard shell
- feed with a beak
- keep a warm body temperature

Mammals

- have a body covered with hair
- keep a warm body temperature
- have live young fed on milk from mammary glands
- have external ears

The five groups of vertebrates

Some important groups of invertebrates

Jellyfish (e.g. jellyfish, sea anemone)

- are symmetrical all round
- live in water
- have a mouth surrounded by stinging tentacles

Segmented worms (e.g. earthworm, leech)

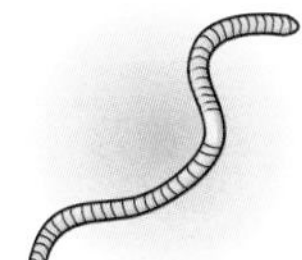

- have a body divided into many similar segments

Arthropods

- have a body with a hard exoskeleton
- have a body in segments
- have many pairs of jointed limbs

Molluscs (e.g. snail, slug, mussel)

- have an internal or external shell
- move using one muscular foot

Cephalopods (e.g. octopus, squid)

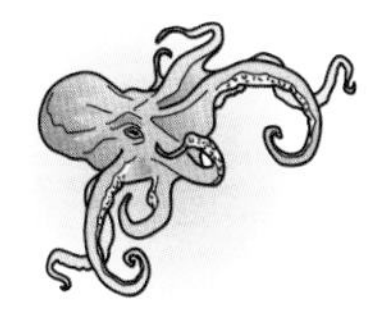

- have a well-developed head and eyes
- use tentacles for catching prey
- move by squirting water out of the body

Starfish (e.g. starfish, sea urchin)

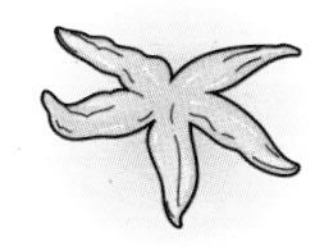

- are symmetrical all round
- have spiny skins
- have a hard outer skeleton with tube feet

The four groups of arthropods

Insects (e.g. butterfly, dragonfly)

- have three parts to the body
- have three pairs of legs
- have two pairs of wings

Spiders (e.g. tarantula, scorpion)

- have two parts to the body
- have four pairs of legs

Crustacea (e.g. prawn, crab, woodlouse)

- have many parts to the body
- have ten pairs of legs

Myriapods (e.g. millipede, centipede)

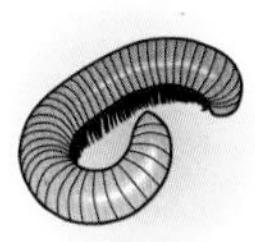

- have a body completely in segments
- have one or two pairs of legs on each segment

Sorting animals

- Sort some pictures of animals into vertebrates and invertebrates.
- Sort each pile into smaller groups.
- Write a label card with the reasons for your decision.

SUMMARY QUESTIONS

1 ☆ Which features do all vertebrates share?

2 ☆☆ You are shown an animal with scales. Which groups could it belong to?

3 ☆☆ Look at the woman with a pushchair. Which group does she belong to?

Key words

invertebrate
sorting
vertebrate

Sorting plants

LEARN ABOUT
- the main groups of plants
- using keys

Sorting plants

Plants are living organisms that make their own food using **chlorophyll** and light energy.

ICT CHALLENGE

Find out about one plant we use for food or a source of useful materials. Make a poster about your plant.

Sorting and using a key

- Look at each of the plants your teacher has provided.
- Use the descriptions in the pictures of plant groups to help you decide which kind of plant each is.
- Use a key to help you identify one specimen.

There are two main groups of plants. **Flowering plants**, for example daffodils, have flowers and reproduce using **seeds**. Most of the plants you see every day are flowering plants. **Non-flowering plants**, such as ferns, do not have flowers. They reproduce using **spores** or specialised sex cells. There are several groups of non-flowering plants.

Q1 Use the guide to decide which group each of these plants belongs to.

Mosses

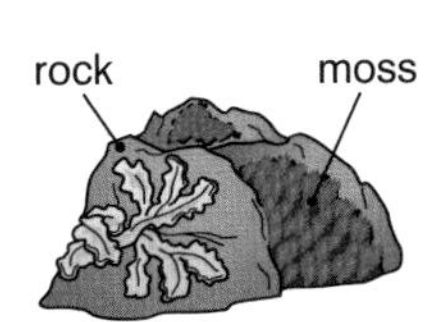

- have no true roots
- have stems or leaves
- produce spores
- are usually small and grow best in damp conditions

Ferns

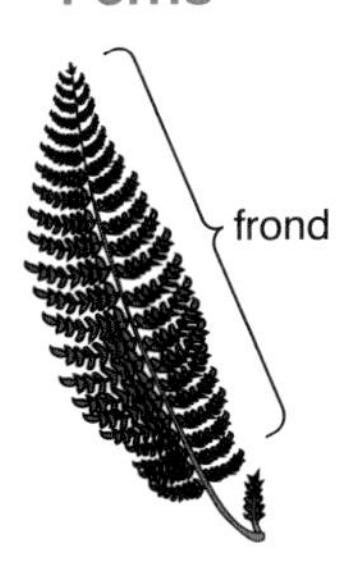

- have large fronds
- grow from a stem with roots
- reproduce by spores produced under the leaves

Conifers

- are trees
- usually keep their leaves during winter
- have needle-shaped leaves
- reproduce by seeds carried in cones

Flowering plants

- reproduce using flowers
- have seeds are protected inside a fruit

The main plant groups

Using keys

Scientists use a **key** to help identify plants and animals. Each species of animal or plant has its own particular combination of features. A key uses animal or plant features to decide which group an unknown specimen belongs to, and then which species it is.

Answering questions leads you to smaller and smaller groups that share the same features.

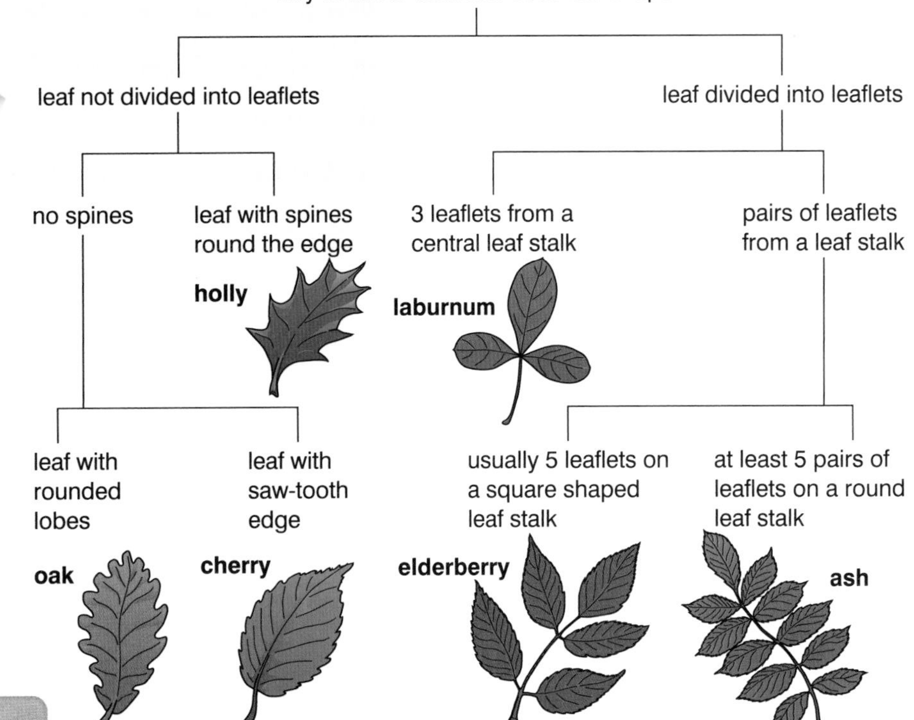

SUMMARY QUESTION

1 ☆☆ Reese found a plant growing near the hedge at the back of the school. She read this description of it in a guide to plants. Use the pictures above to decide which group it belongs to.

Specimen Z

Common everywhere in woods. The fronds grow in a tuft to 1 metre high from a central point. It has extensive roots. Fully grown fronds have large triangular blades, each heavily divided into lobes. Spores develop in a row under the lobes of the frond.

Key words

chlorophyll
conifer
flowering plant
key
seed
spore

LEARN ABOUT
- how we vary

Spotting differences

Cattle all around the world are members of the same species, *Bos taurus*, but they don't look the same everywhere. As you can see, these cattle have horns of different lengths and shape. Differences between one individual and another are examples of **variation** in a species.

Ankole cattle
an African breed

Zebu
a breed found in India

Jersey
bred for a rich creamy milk

Hereford
bred to provide beef

Domestic cattle from around the world

Q1 Can you see any other differences?

The differences between the cattle are easier to spot than the differences between zebras, meerkats or penguins. They recognise each other by very small differences in voice, smell or appearance.

Meerkats keeping a sharp lookout for predators

I wish we were a bit more identical
1
2

How do we vary?

People differ from each other in many ways. We have different eye colours and hair texture. Some of us are tall, some small, and some are in between. Some of us are male and some female. We all have different fingerprints.

Male or female is a simple either/or choice. You belong to one category or another. This sort of variation is called **discontinuous variation**.

However, if we measure how tall Year 7 pupils are we find that people do not fall into distinct categories. We get a range of values. Each person's height is somewhere between about 135 cm and 175 cm. This is described as **continuous variation**.

This sort of feature is affected by the **environment** an individual grows in. For example malnourished children are often smaller than well-fed children. Trees growing in windy places are often stunted.

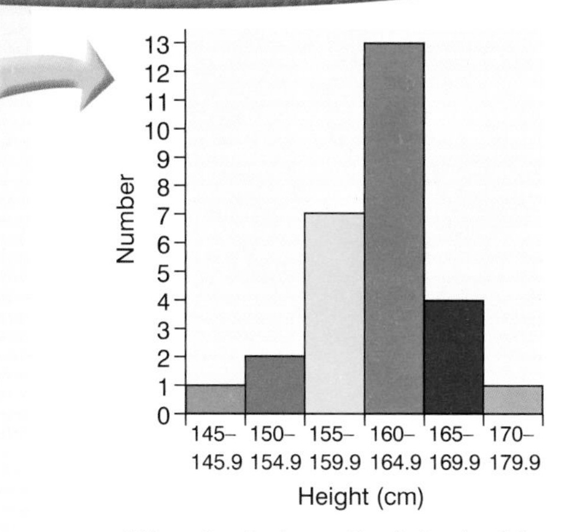

The height of girls in Year 9 shows continuous variation

Inheriting features

Where do these differences come from? Some are **inherited** from our parents. The shape of your nose, the gap between your front teeth, what sex you are and your hair type are inherited from your parents. The cattle on the right have inherited the ability to give large amounts of milk, or to be strong enough to pull a plough or a cart.

All animals and plants inherit features from their parents. Animals or plants that have inherited useful variations will cope well with the problems of their environment.

Investigating variation

- Dogs at a dog show belong to the same species. Use pictures of different varieties of dog to describe ways in which the breeds differ from each other.

SUMMARY QUESTIONS

1 ☆ Look at the photo.
 a) List the differences that you can see between the two individuals.
 b) Find two features that you think they inherited.

Brothers

2 ☆☆ Look at the pictures of cattle on the opposite page.
 a) Choose three features that vary and draw up a table of differences.
 b) Which of these features do you think the cattle inherited?

Key words

continuous variation
discontinuous variation
environment
inherit

Learn about
- genes
- how we inherit features

Inheritance

To have blue eyes or brown eyes we need information for making the colour. This sort of information is carried as **genes**. Every cell of our body has genes that carry the information needed to build a human being.

Genes are found in the nucleus of cells. They are arranged on structures called **chromosomes** in the nucleus. Genes are made of a substance called **DNA** – **d**eoxyribo**n**ucleic **a**cid. It is the longest molecule in living things.

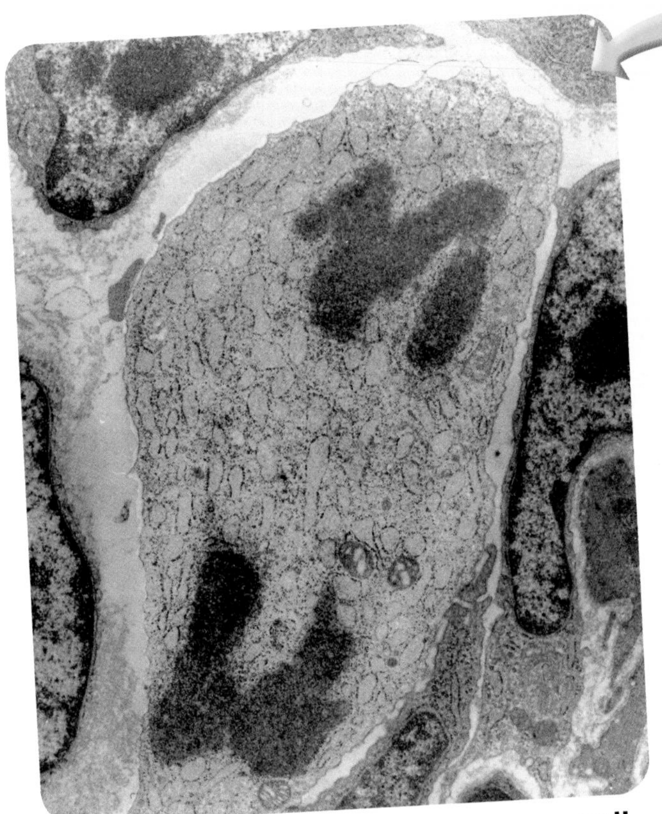

We can see chromosomes when a cell is dividing to make two new cells

What sort of information do genes carry?

Genes are responsible for many of the features of living organisms. Genes are responsible for your blood group, your eye colour, the size of your nose and the shape of your feet. They are also responsible for a lot of animal behaviour, such as hunting and mating.

The largest length of DNA discovered so far is in the tiny *Amoeba dubia*, which is only 0.4 mm long.

How do you get your genes?

Children look like their parents because they have **inherited** their parents' genes. When men and women make sperm and eggs, copies of their genes are passed into the sperm and egg cells. Plants pass copies of their genes into pollen and ovules.

Sperm and egg cells carry copies of half of each parent's genes. When an egg is fertilised by a sperm, or pollen fertilises an ovule, the two half sets of genes join together to make a full set.

Which features seem to be inherited in this family?

A child inherits genes from both parents. It will not be exactly like either parent, but has a combination of features from each of them. Brothers and sisters look similar because they have inherited similar combinations of genes from their parents.

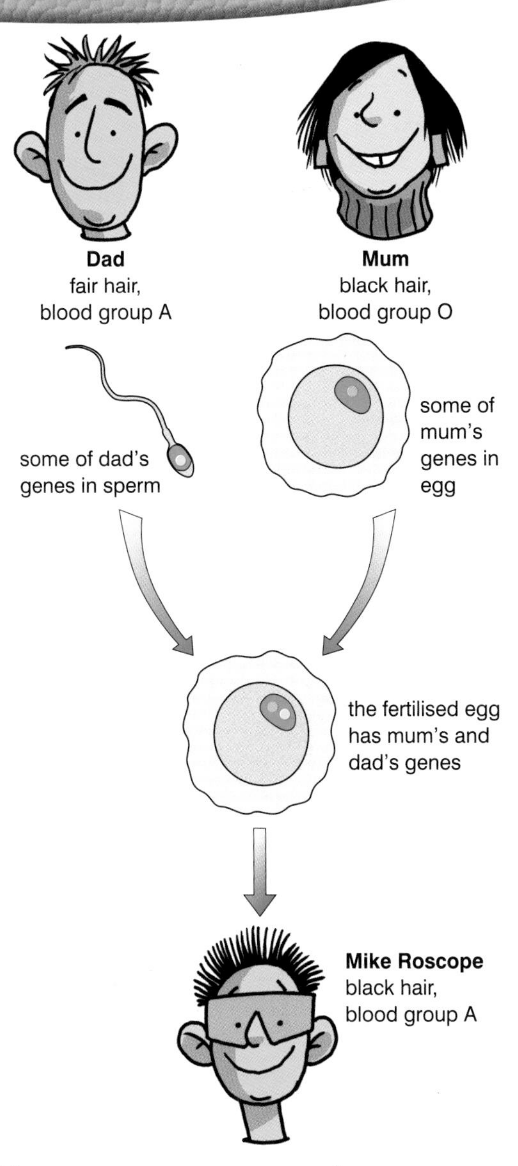

How genes are passed on

Studying inheritance

- Look at some family photos – your own or a friend's. Look for features that have been inherited in the family.

SUMMARY QUESTIONS

1. ☆ Explain why some people in your group have a different eye colour from you.
2. ☆☆ Think back to the unit on reproduction (Unit 7B). Why do identical twins have fewer differences than other brothers and sisters?

Key words

chromosome
DNA
gene
inherit

Making use of genes

LEARN ABOUT
- selective breeding

Wild ancestors

People have been keeping cattle for about 6000 years. The early farmers tamed wild aurochs. The breeds we have today are descended from them. All our farm animals and crops were bred from wild ancestors.

All of our variously coloured hamsters are bred from a single female golden-brown hamster found in the wild in the 1930s.

Breeding animals

Farmers in Africa, where it is very hot, don't want the same sort of cows as farmers in cool, damp northern Europe. Farmers need animals that can cope with the climate where they live. The animals must be able to thrive on the local vegetation and resist local diseases.

Farmers keep their most successful animals to breed from. For example, farmers select cows that give plenty of milk, or chickens that lay plenty of eggs. Often these features are controlled by **genes**.

N'Dama
resists African sleeping sickness infection

Highland
long hair helps survive harsh winters

Texas Longhorn
grows well on poor, dry vegetation

Cattle adapted to different environments

The selected animals are the parents of the next generation. Their young are likely to inherit their good features. The farmer carries on breeding from the best of every generation. In this way, the farmer's animals are gradually improved. This is called **selective breeding**.

Crops and flowers are bred in the same way. The growers keep bigger or better seeds and fruits for planting. Selective breeding has produced wheat from a wild grass and pineapples from a spiky bush.

The breeding process is also used to meet new demands. For example, fruit-growers want better colours or flavour for their fruit, fewer pips, or a more suitable size for school lunch boxes.

Breeding sweet oranges began in China in the Iron Age. They were brought along caravan routes with limes to Europe. Lemons and grapefruit were bred later.

ICT CHALLENGE

Find out about a rare breed of farm animal such as a Gloucester Old Spot pig or White Park cattle.

Breeding

- Use a computer simulation to carry out a breeding experiment.

SUMMARY QUESTIONS

1 ☆ Explain how the information for good features is carried from parent to offspring.

2 ☆☆ Which features might a farmer look for in a dairy cow?

Key words

breeding
genes
selective breeding

IDEAS AND EVIDENCE

Conserving genes

A lot of hard work goes into **breeding** new, more productive varieties of crops and farm animals. Growers quickly switch to the new varieties. Before long, the older varieties become very rare.

We can lose features altogether through breeding. For example, a breeder selecting flowers for a clear red colour may choose plants that have a good colour but don't have a very strong scent. As a result, the final **variety** may have only a faint scent.

Purple coloured potatoes have become very rare

We must keep wild plants and old varieties going. They act as a store of the features that have been lost in the new varieties – and these may be important in the future.

The human population is growing so fast that we have to grow crops in drier and poorer soils. By using older varieties in breeding programmes, breeders can get back the genes that the plant needs to grow well in poor conditions.

A **gene bank** keeps seeds from older plant varieties. They are dried and stored in a deep-freezer at −20°C. Every few years some plants are grown to provide fresh seeds. There are gene banks for rare breeds of farm animals too.

DANGER! AVOID THESE COMMON ERRORS

We can only inherit features that are controlled by genes. We cannot pass on features that we have acquired during our lives to our children. For example, you may have your ears pierced or have a large waistline – but your children won't be born with pierced ears or needing big trousers!

Key words

gene bank
selective breeding
variety

UNIT REVIEW

REVIEW QUESTIONS

Understanding and applying concepts

1 **a** *Triturus cristatus* is a vertebrate with a moist skin without scales. Which group does it belong to?

b Which features would best help you to decide if an interesting living creature you had found was an insect?

c *Lacerta agilis* is a vertebrate that lays eggs and has a body covered in scales. Which group does it belong to?

d *Taraxacum officinale* has true leaves and roots and produces seeds. Which group does it belong to?

e *Triturus cristatus*, *Lacerta agilis* and *Taraxacum officinale* are scientific names. Use a guide to British animals and plants to find out the common names of these species. (7)

2 The following animals are all arthropods. Sort them into insects, arachnids, myriapods and crustacea.

woodlouse bee ant ladybird wolf spider red admiral centipede scorpion lobster brine shrimp daddy-long-legs stag beetle house fly crab garden cross spider wasp dragonfly grasshopper froghopper mosquito tarantula pond skater millipede

3 Copy and complete the sentences, using words from the list below.

Inherited features are carried as .... They are arranged on ... that are found in the ... of a cell. Everyone is different, because we have different combinations of inherited features. We inherit our features from our ... and mother. They pass on these features in sperm and ....

Farmers choose animals and plants with good features for breeding. The young that have inherited good features from both ... are kept. This process is called ....

nucleus selective breeding eggs genes father parents chromosomes

4 Mr Williams loved the fuchsias in his garden. He wanted more, so he chose his best plant and cut off some shoots for cuttings. He planted them in pots and put them in a sheltered, shady part of the garden to root and grow. He expected them all to grow the same. Three months later, one was 10 cm tall with four leaves, another 15 cm with six leaves, and one 23 cm tall with seven leaves.

a Why did he expect all his cuttings to be the same?

b Why do you think they were different? (2)

5 Make a 'spider' diagram of animal classification. Make sure you include all these words in your map:

animal vertebrate invertebrate arthropod insect wings

SAT-STYLE QUESTIONS

1 Here are the five groups of vertebrate animals.

fish amphibians reptiles birds mammals

a Give one way that birds are different from the other vertebrate groups

b Which groups have scales?

c Some of the sentences below are true and some are false. Put (T) for true or (F) for false beside each sentence.

all vertebrates have a backbone
all vertebrates have warm blood
all vertebrates give birth to live young
all vertebrates have four limbs
all vertebrates have a tail (7)

2 Jasmine and Meena are identical twins but they are in different classes at school. Jasmine has been preparing a presentation in ICT. Meena has been doing dance.

Here are four sentences describing a twin.
She is breathing fast
She has dark brown hair
Her skin is flushed and redder than usual
She has brown eyes

- **a** Which two sentences in the list above could describe **both** Jasmine and Meena?
- **b** People who do not know the twins very well often muddle them up because they look so similar. Choose the correct explanation from the sentences below.
  - **i)** Their parents brought them up together and made sure they both had the same food and the same things.
  - **ii)** They inherited the same genes from their parents. (3)

3 Use the diagrams and your own knowledge to fill in the table below:

Characteristic	Animal A	Animal B	Animal C
number of legs			
number of wings			
body divided into segments			
eyes			
antennae			

(5)

4 Horse racing is a multi-million pound business. Champion racehorses could pass on their winner's genes to their offspring. Winning horses are fast, but they need other features too. They have to cope with the challenges of races, travelling long distances around the country, and being ridden by several different jockeys.

Think of six features that a breeder might look for when breeding a future winner and explain why they are important. (6)

A winning racehorse is worth millions for its breeding potential

5 Class 7D measured their hand span. They stretched their fingers out and measured the distance from the tip of the thumb to the tip of the little finger.

There were 27 people in school that day. Here are the data they collected.

Hand span (cm)

21.0	20.0	17.5	19.5	20.5
19.3	19.0	20.5	18.2	19.5
21.3	20.2	22.3	21.0	20.8
21.9	18.6	19.2	22.8	18.3
19.5	21.4	18.8	18.8	22.0
20.5	20.4			

- **a** Group the data ready to plot a bar chart. Use the size ranges of 17.0–17.9 cm, 18.0–18.9 cm, and so on.
- **b** Plot a bar chart of your data.
- **c** What was the most frequent size of hand span?
- **d** What was the largest hand span? What was the smallest?
- **e** Why do you think there is such a variation?
- **f** Is this continuous or discontinuous variation? (10)

Key words

Unscramble these:
frenioc
the rini
raviotina
ever trable
escipes

7E Acids and alkalis

Coming soon

What's it all about?

Acids have a really bad reputation with most people. They only think of acids as dangerous, fuming liquids.

However, we all come across acids every day. For example, oranges, apples, lemons, yoghurt, tea and vinegar all contain weak acids.

In this unit we look at a range of **acids**. We also look at their chemical opposites, called **alkalis**. We will find out how we make use of them. We will also see how to test a solution to find how strongly or weakly acidic it is.

What do you remember?

You already know about:

- some substances we classify as acids
- solids that dissolve to form solutions
- mixing some substances that change to make new substances

1 Which word describes a solid that can dissolve in a liquid?

solvable dissolvable soluble molten

2 Which of these substances does NOT dissolve in water?

instant coffee sand salt sugar

3 Which of these pairs will form new substances if you mix them together?

bicarbonate of soda and vinegar
milk and water sand and flour
sugar and water

Ideas about acids

QUESTIONS

Look at the cartoons above:
Now discuss these questions with your partner.

a) Have you ever used any acids – at home or at school? Name any acids that you know.

b) Do you think that all acidic solutions are dangerous? Do you think that they all burn holes in whatever they touch? What evidence have you got for your ideas?

c) How will you stay safe doing experiments with acids at school?

Acids all around

LEARN ABOUT

- acids that we use at home
- not all acids are hazardous

Acid rain

You've probably heard of acid rain.

The gases that cause acid rain come from cars, power stations and factories. However, rain has always been naturally **acidic** – but only slightly. That's because one of the gases in the air, **carbon dioxide**, **dissolves** in water. It forms a weakly acidic **solution** of **carbonic acid**.

Q1 Is rain the purest type of water you can get? Why?

Q2 Why you think that acid rain is worse in some places?

Label from a fizzy cola drink

Acids in drinks

The acids in acid rain are a problem for us, but many acids are useful. Weak solutions of acids have a sharp taste. You will recognise the sharp, tangy taste of vinegar. It contains **ethanoic acid**. Fruits such as oranges, lemons and limes also contain acids.

The food industry uses acids in many products. Look at the label from a fizzy drink.

The 'fizz' in a fizzy drink comes from carbon dioxide gas. But there is a lot more gas dissolved in the drink than there is in rain!

Q3 How do you think that makers of drinks get the 'fizz' in their cola?

Q4 Which other acid is in the cola above?

I THINK WE SHOULD CUT DOWN ON THE ACID, BOYS...

Fizzy drinks

- Design a safe test you could do to work out which drink has more fizz.
- If you can, think of some way to measure the difference. Your teacher might give you a sheet with some ideas for you to try.
- Show your plan to your teacher before you try out your test.

Acids in foods

Drinks aren't the only things that have acids added to them before we buy them. For example, E300 is ascorbic acid, which you know as vitamin C. We add this to flour.

INGREDIENTS	INGREDIENTS	INGREDIENTS
Maize, sugar, malt flavouring, niacin, iron, vitamin B6, riboflavin (B2), thiamine (B1), folic acid, vitamin B12.	Sugar, vegetable oils, wheat flour, dried skimmed milk, lactose, butter oil, maltodextrin, flavourings, emulsifier (lecithin), raising agents, (sodium bicarbonate, ammonium bicarbonate, tartaric acid), salt, colour (annatto).	Water, blackcurrant juice, citric acid, sweeteners, vitamins (C, niacin (which is nicotinic acid), pantothenic acid, B6, B12, D) acidity regulator, preservatives.
Breakfast cereal	Chocolate biscuit	Blackcurrant vitamin drink

Food labels

Citrus fruit contains ascorbic acid

Food survey

- Do a survey of food and drink labels. Find as many acids as you can.
- Try to find out why the acid has been added.

Gruesome science

Supermarkets are thinking of ways to dissolve carbon dioxide gas into fruits. They hope this will get more children eating fruits. Tests show that the fizzy fruits taste stronger – but bananas explode!

SUMMARY QUESTION

1 ☆ Copy and complete the sentences, using words from the list below.

hazardous weak carbonic dioxide ethanoic

Acids are all around us in everyday life. For example, carbon . . . gas dissolves in rain to form . . . acid (a . . . acid). We find acids in many foods, such as . . . acid in vinegar. These acids are not . . . .

Key words

acidic
carbon dioxide
carbonic acid
dissolve
ethanoic acid
solution

Learn about

- common hazard symbols
- how to deal with acids or alkalis if you spill them

What are hazard symbols?

Sometimes in Science lessons you will use substances that are **hazardous**. You need to know about the dangers and make sure you use them safely. Your teacher will advise you, of course!

Here are some hazard signs you might come across on the labels of chemicals.

CORROSIVE
These substances attack and destroy living tissues, including eyes and skin

IRRITANT
These substances are not corrosive but can cause reddening or blistering of the skin

HARMFUL
These substances are similar to toxic substances but less dangerous.

HIGHLY FLAMMABLE
These substances catch fire easily

OXIDISING
These substances provide oxygen, which allows other materials to burn more fiercely

TOXIC
These substances can cause death. They may have their poisonous effects when swallowed, or breathed in, or absorbed through the skin

Which hazard symbol do you see on signs at a petrol station?

harmful   highly flammable
corrosive   oxidising agent

Have you ever noticed the hazard warning symbols on any products in your home?

Look at the examples below.

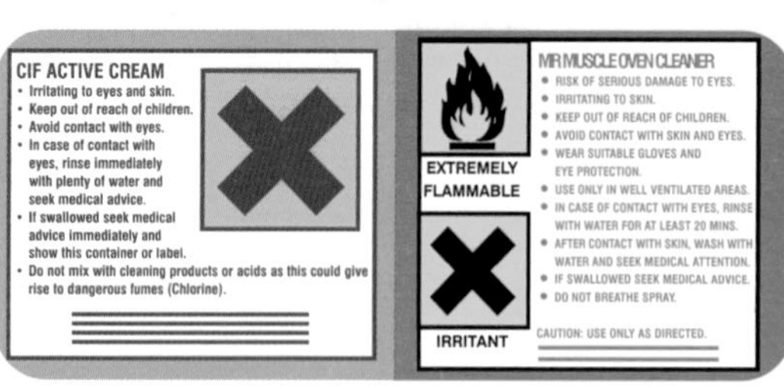

Hazard symbols on everyday products

Q2 Which of these would be most dangerous to get on your skin?

washing-up liquid
oven cleaner
vinegar
lemon juice

Rubber gloves are enough for most people, Mike.

Gruesome science

Accidents with alkalis are often more serious than those with acids. Alkalis attack fats and oils in your body tissues and eyes. They also attack other substances, and the damage can't be reversed.

Concentrated and dilute acids

- Watch your teacher drop **dilute** sulphuric acid on some sugar.
- Now watch as your teacher adds **concentrated** sulphuric acid to sugar.
  a) Describe the differences you see.
  b) Which of the two bottles of acid is more hazardous to handle? Why?

SAFETY: Usually, we can make an acidic solution less hazardous by adding water to it. In other words, we **dilute** it. However, adding water to concentrated sulphuric acid is very dangerous. To make dilute sulphuric acid, we must always add concentrated acid to water.

Gruesome science

Derbyshire council is spending £75 000 finding out the damage that dogs do to its lamp posts. They are worried that acid in the dog's urine could react with the concrete at the base of street lights and weaken it.

Transporting hazardous loads

You've probably seen the hazard signs on the back of tanker lorries. These help the emergency services if there is an accident. The codes tell fire fighters how to deal with the hazards.

Look at this hazard sign on this tanker lorry.

Q3 What does the picture on the sign tell you?

SUMMARY QUESTIONS

1 ☆ Write a list of safety rules for a Year 7 class who are doing experiments with acids and **alkalis**. The rules should include what to do if some acid or alkali spills on a bench.

2 ☆☆ Do a survey of household products. Record any hazard symbols you find. Put your results in a table.

Key words

alkali
concentrated
corrosive
dilute
hazardous
irritant

LEARN ABOUT
- making indicators
- using universal indicator and the pH scale

What are indicators?

Some substances change colour in acid and alkali. These are called **indicator**s. Dyes extracted from plants often make good indicators. You can make your own indicator in the next experiment.

You can try making an indicator from flower petals

There must be a way to tell which of these acidic solutions is corrosive, which is irritant and which is so dilute it is harmless.

Making an indicator

STEP 1

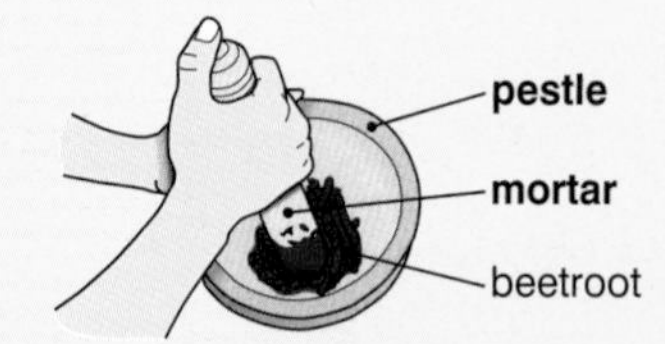

Chop up some red cabbage, or beetroot. Place it in a mortar and pestle. Add a little water, then crush it.

STEP 2

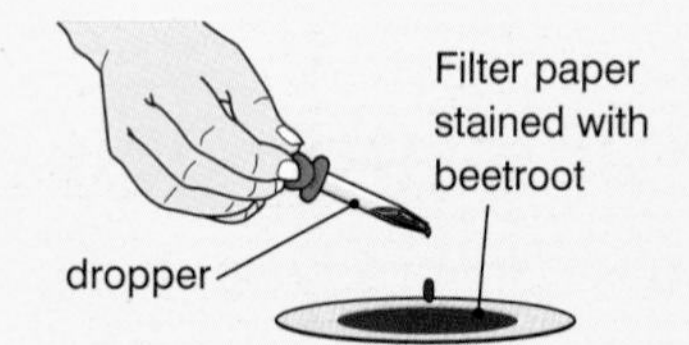

Use a dropper to put the dye onto a piece of filter paper

STEP 3

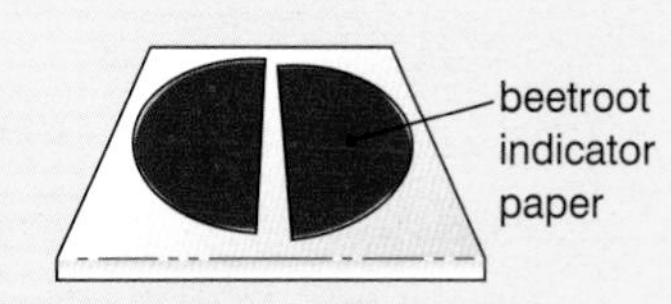

Leave the stained filter paper to dry in a warm place. Once it has dried, tear it in half. Place each half on a white tile. You have now made indicator paper.

STEP 4

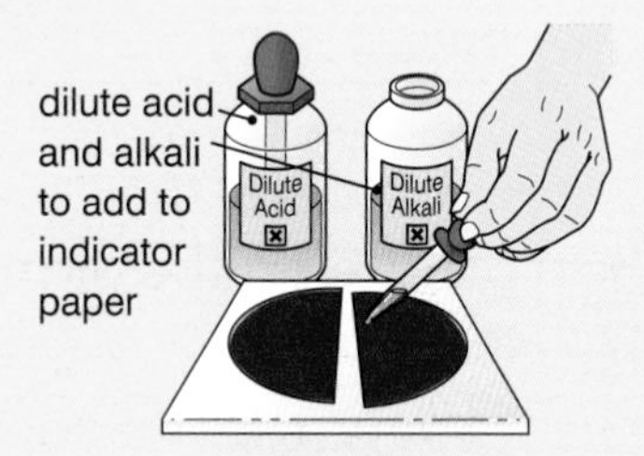

Add a few drops of acid to one half of your indicator paper. Add a few drops of alkali to the other half.

What colour does your indicator turn when you put it in:
a) acid? **b)** alkali?

The pH scale

Universal indicator tells us how strongly acidic or alkaline solutions are. It contains a mixture of dyes and so it can turn a whole range of colours. It can even tell us if a solution is neither acidic nor alkaline. We call these solutions **neutral**. In a neutral solution universal indicator turns green.

We match the colour of the universal indicator to a **pH number**. These numbers are shown on the **pH scale**.

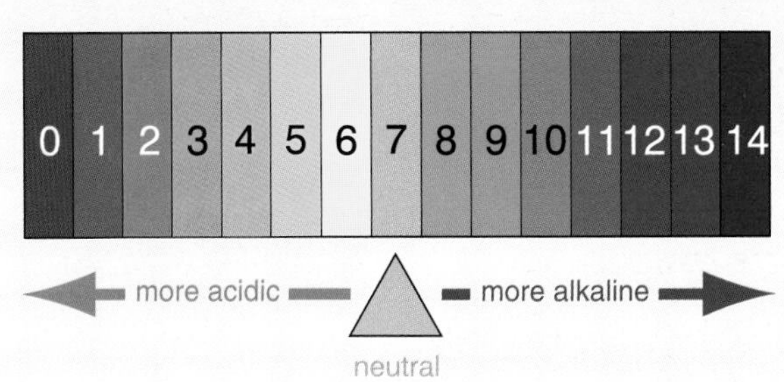

The pH scale

Q1 Which of these pH numbers indicates the most strongly acidic solution?

1 5 7 14

Testing the pH of solutions

- Test the pH of different solutions.
- Add a few drops of universal indicator to a small volume of each solution in separate test tubes.
- Show your results in a table like this one.

Solution	Colour of the universal indicator	pH number of the solution	Acidic, alkaline or neutral

a) Which is the most strongly acidic solution you tested?
b) Which is the most strongly alkaline solution you tested?
c) Name any solutions you tested that are neutral.

SUMMARY QUESTION

1 ☆ **a)** Arrange these solutions in order, with the most alkaline first:
pH numbers: **2 8 6 3 12 7**
b) Which of the solutions in part (**a**) is neutral?
c) What colour is universal indicator in a neutral solution?

Key words

indicator
mortar
universal indicator
neutral
pestle
pH number
pH scale

7E4 Reacting acids with alkalis

LEARN ABOUT
- using acids and alkalis in everyday life
- how pH changes when acid is added to alkali

At the base of each hair on a nettle leaf, there is a store of venom. It contains methanoic acid.

What is a nettle sting?

Have you ever been stung by nettles? You usually find dock leaves growing near nettles. When you rub these on the nettle sting, they soothe the pain.

The nettle sting contains an acid. One theory says that the dock leaf releases a weak alkali that 'neutralises' the acid in the sting.

In this lesson you can find out more about the chemical reaction that we call **neutralisation**.

Adding acid to alkali

- Put 5 cm^3 of dilute **sodium hydroxide** solution in a test tube.
- Place the test tube in a test-tube rack. Then add a few drops of universal indicator solution.
  - **a)** What is the pH number of the solution?

dilute acid
dilute alkali plus universal indicator

- Add 4 cm^3 of dilute hydrochloric acid to the same test tube.
  - **b)** What is the pH number of the solution now?
  - **c)** What does this tell you about the solution?
- Now use a dropping pipette to add more acid to the test tube, a drop at a time. Try to stop the reaction just as the universal indicator turns green. You need to be very careful to do this.
  - **d)** What happens to the pH of the solution if you add too much acid?
  - **e)** How can you make your solution neutral without starting all over again?

Come on, Pete... you must have managed to neutralise that solution by now!

A neutralisation reaction

We can think of acids and alkalis as chemical opposites. They react together and 'cancel each other out'. When you mix them in equal amounts, they form a neutral solution. That's why we call the reaction of an acid with an alkali a **neutralisation reaction**.

Energy changes

We can get a neutral solution from an acid and an alkali. This shows that a chemical reaction takes place. A new substance or substances must be formed. If mixed in the right amounts we get:

acid + alkali → neutral solution

Also during chemical reactions, energy is given out or taken in. This means that the solution will get hotter or cool down.

Try the ICT CHALLENGE. You can follow the temperature change as the neutralisation reaction happens.

ICT CHALLENGE

Try the 'temperature change' experiment or watch your teacher do it. A **temperature sensor** and **data logger** will monitor the temperature change as the reaction is taking place. Your teacher will give you a sheet to follow.

Useful acids and alkalis

- Do some research into acids and alkalis and how we use them. You can use books, videos, leaflets, CD-ROMs or the Internet.
- Here are some key words to help you in your search. Choose one of them to use. Present your work as part of a class folder called 'Uses of acids and alkalis'.

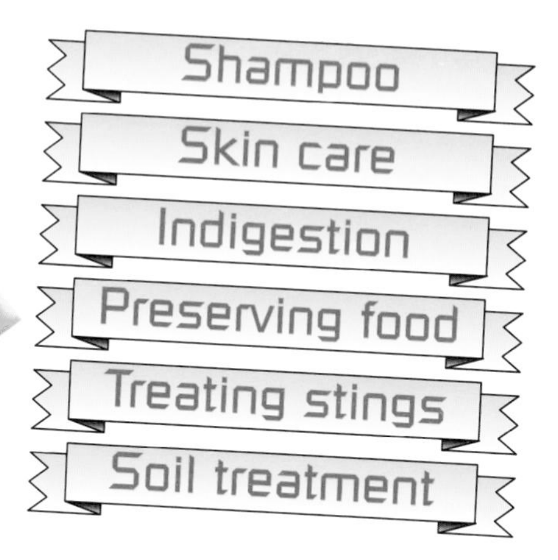

SUMMARY QUESTION

1 ☆ Copy and complete the sentences, using words from the list below.

neutral 7 react neutralisation lower

When we add an acid to an alkali, the pH of the solution gets . . . .

When the right amount of acid and alkali . . . together, we get a . . . solution formed. Its pH number will be . . . .

We call this a . . . reaction.

Key words

data logger
methanoic acid
neutralisation
sodium hydroxide
sensor

Investigating neutralisation

LEARN ABOUT
- monitoring a reaction
- carrying out a fair test on indigestion remedies

Using neutralisation

You know that neutralisation is an important reaction.
We use neutralisation in lots of ways in everyday life.
For example, farmers neutralise acidic soils.
We also neutralise acids when we brush our teeth.

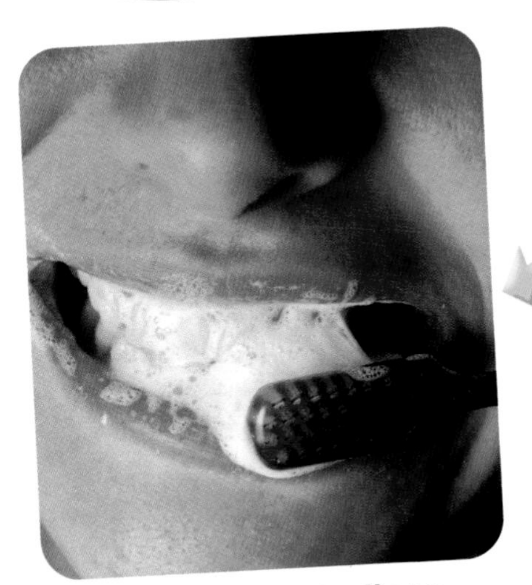

Toothpaste neutralises the acid in your mouth. Acid attacks the enamel on your teeth. It causes cavities.

Q1 Why are toothpastes only weakly alkaline?
Is it because strong alkalis are too expensive?
Or is it because strong alkalis can be corrosive?

Monitoring pH change

You can follow the pH changes as you add an alkali to an acid. You or your teacher will use a **pH sensor**. This will monitor the pH changes. We use a **burette** to add small volumes of alkali to the acid. This experiment is called a **titration**.

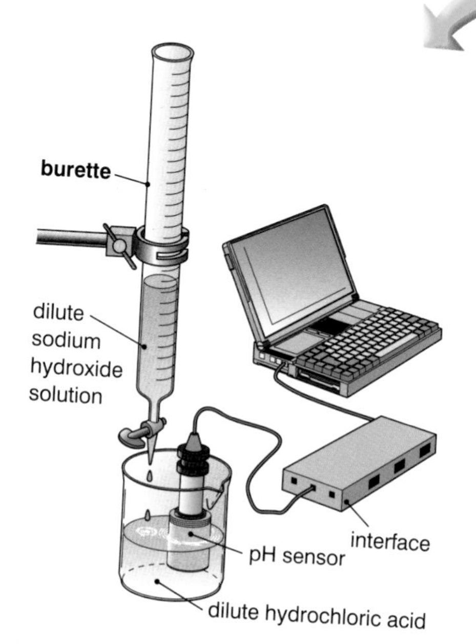

- Put 20 cm^3 of dilute hydrochloric acid in a small beaker.
- Record its pH using a pH sensor.
- Add dilute sodium hydroxide solution to the burette.
  a) Predict how the pH will change as you add this alkali to the hydrochloric acid.
- Now add the sodium hydroxide, 1 cm^3 at a time, to the acid.
- Stir after you add each 1 cm^3 of alkali. Then take the pH of the solution in the beaker.
- You can record your results in a table. Then show them on a graph. Or you can use the computer to do the job for you.
  b) Was your prediction right or not?

Indigestion remedies

Did you know that your stomach contains hydrochloric acid? It helps to break down your food and to kill bacteria.

But sometimes you can produce too much acid. That's when you get that burning feeling we call **indigestion**.

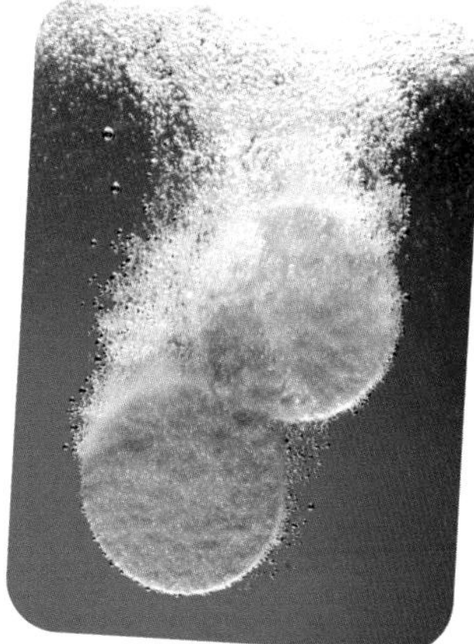

Why don't antacids contain sodium hydroxide?

People can take a tablet to ease the pain. These indigestion remedies are called **antacids**.

Please Molly, I need neutralising... quickly!

All antacids contain substances that react with acids to neutralise them. These substances are called **bases**.

Bases include alkalis, but not all bases dissolve in water. All alkalis do dissolve in water.

Investigating indigestion remedies

SAFETY

You can now test several different indigestion remedies.

- As a group, think of a question you would like to investigate. For example: *Which antacid is best at neutralising acid?*
- Plan your investigation. Remember to make it a fair and safe test.

I don't want to be around when that gas escapes!

Your teacher will give you a sheet to help you plan and write up the investigation. Make sure your plan is checked before you start.

Gruesome science

When you vomit, you get a burning feeling in your throat. This is because hydrochloric acid in your stomach has a pH number of between 2 and 3. The wall of your stomach is lined with mucus to protect it from attack by the acid.

SUMMARY QUESTION

1 ☆ a) What acid causes indigestion?
b) How do we treat the symptoms of indigestion?
c) Design an advert for the best antacid tested in your investigation.

Key words

antacid
bases
burette
indigestion
pH sensor
titration

IDEAS AND EVIDENCE

Acids for health

Have you ever looked at what's in a bottle of vitamin tablets? You will see quite a few different acids listed on the label. The chemical name for vitamin C is ascorbic acid.

The vitamin C is usually in the form of a neutral substance similar to ascorbic acid.

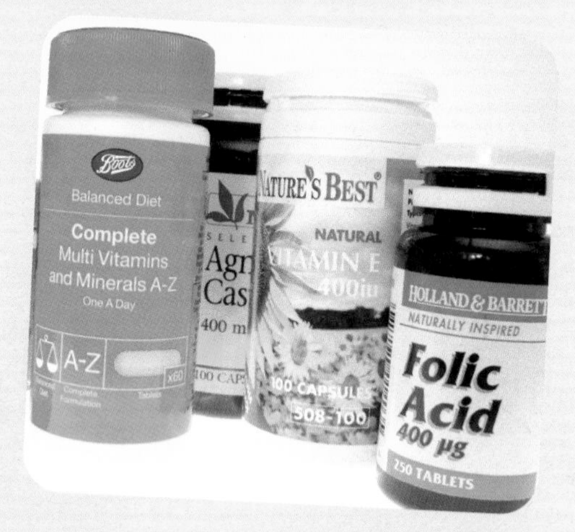

The Ancient Greek doctor, Hippocrates, gave women tea made from the bark and leaves of the willow. It relieved pain during childbirth. In the 1800s, we discovered that the painkiller from the willow was salicylic acid.

You will also find folic acid in most multi-vitamin tablets. This vitamin is essential during pregnancy. It can also reduce the risk of heart disease. Pantothenic acid is also known as vitamin B5. It is another ingredient in vitamin tablets.

This pregnant woman is taking a tablet containing folic acid supplement

Nowadays we make aspirin from salicylic acid

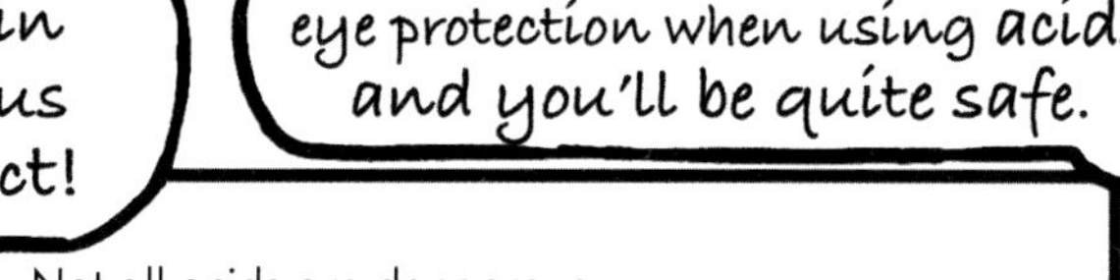

Always behave sensibly and wear eye protection when using acids and you'll be quite safe.

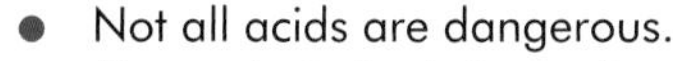

Well, there's ascorbic acid in oranges, ethanoic acid in vinegar, citric acid in citrus fruit... loads of acids in fact!
So my shampoo is weakly acidic with a pH of 5.5!
Universal Indicator will tell me which is which!
Don't worry! I'll mop it up with plenty of water... it is not concentrated sulphuric acid, but we still have to take care.

A little vinegar is harmless... the ethanoic acid in it is only a weak acid.

- Not all acids are dangerous.
- Concentrated solutions of strong acids are corrosive. They attack materials and living tissue.
- The more **dilute** you make an acid (by watering it down), the safer it is to use.
- Indicators are substances that change colour in acids and alkalis.
- Universal indicator changes through a range of colours.
  We can match the colour to a pH number on the pH scale:

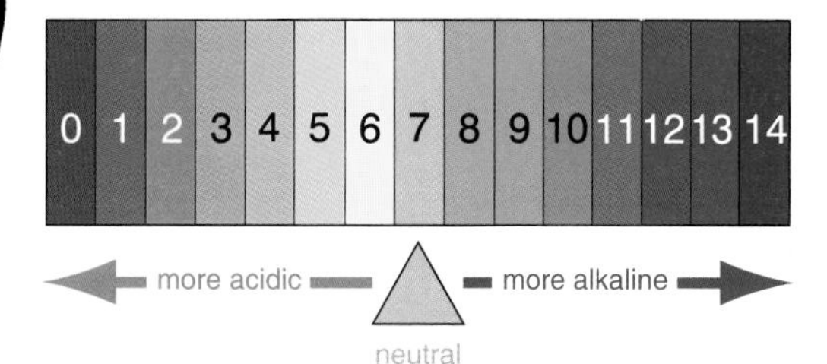

- Acids and alkalis react together in a neutralisation reaction.

We use **acids** and **alkalis** in everyday life, but most solutions are not **corrosive**.

You have to be careful when using the **pH scale** because the neutral point (neither acidic nor alkaline) is at pH 7. Below pH 7, the *lower* the pH number, the *more* strongly acidic a solution is. Above pH 7, the *higher* the pH number, the *more* strongly alkaline a solution is.

Acids and alkalis react together in a chemical reaction called **neutralisation**. But you only get a neutral solution if you mix *exactly* the right amounts of acid and alkali together.

Key words

acid
alkali
corrosive
dilute
neutralisation
pH scale

UNIT REVIEW

REVIEW QUESTIONS

Applying and understanding concepts

1 Copy and complete the sentences, using words from the list below. Use the summary on page 91 to help you.

diluting neutralisation concentrated colour less neutral higher alkalis alkaline tissue corrosive universal

A . . . solution of a strong acid will attack materials and living . . . .

We call these solutions . . . . We can make them safer to use by . . . them with water.

Indicators change . . . when we add them to acids or . . . .

We use . . . indicator to find out the pH number of a solution.

Solutions with a pH number . . . than 7 are acidic, whereas those . . . than 7 are . . . . A solution with a pH number of 7 is called . . . .

When an acid and alkali react together we call it a . . . reaction.

2 Jason tested some unknown solutions with universal indicator. He recorded the colours in the table.

Solution tested	Colour of universal indicator
A	green
B	red
C	turquoise (blue / green)
D	orange / yellow
E	purple

a Which solution is strongly acidic?
b Which solution is weakly acidic?
c Which solution is strongly alkaline?
d Which solution is weakly alkaline?
e Which solution is neutral?
f Solution B is marked with this hazard sign.

What does this sign tell us?

3 Copy Nicky's concept map below and label the arrows for her.

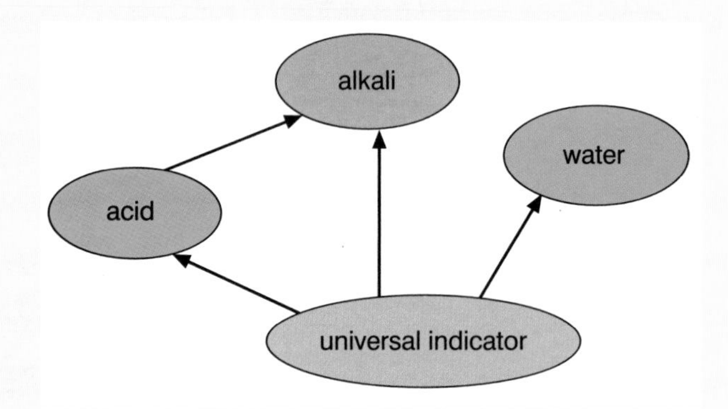

Making more of maths

4 A group of students monitored the pH of an acid as they added alkali to it.

These are their results.

Volume of alkali added (cm^3)	pH number
0	4.0
5	5.0
10	5.5
15	6.0
20	6.5
25	9.0
30	10.0
35	11.0

a Draw a graph of their results.
b How much alkali was added to produce a neutral solution?

Extension questions

5 Make a list of all the acids you have come across in this unit.

6 Do some research to find out the products that we can make from sulphuric acid. Present your findings on a poster to display to the rest of your class.

SAT-STYLE QUESTIONS

1 You can try to make your own indicator using some coloured flower petals.

The first step involves crushing and grinding the petals to release the colour into a little water.

a What apparatus would you use to grind up the petals with water?
. . . and . . . . (2)

A group of students made some indicator solution from three different colours of petal. They added their indicator to an acid and an alkali.

These are their results.

Colour of petals and indicator solution in water (pH 7)	Colour in a solution of pH 1	Colour in a solution of pH 14
yellow	yellow	yellow
red	red	green
purple	pink	blue

b What colour would the red petal indicator be in a solution of sulphuric acid? (1)

c What colour would the purple petal indicator be in a solution of salt which is neutral? (1)

d What colour would the purple petal indicator be in a solution of sodium hydroxide? (1)

e Explain which colour of flower petal would make the best indicator for both acids and alkalis. (1)

2 The table shows the pH numbers of five solutions whose labels have been lost.

Solution	pH number
A	6.0
B	7.5
C	7.0
D	4.5
E	8.0

a Which solutions are acidic? (2)

b Soap solution is weakly alkaline. Which of the solutions could be soap solution? (2)

c i) Give two solutions from the table above that would react together. (1)

ii) What do we call this type of reaction? (1)

3 Nina and Alex tested 3 different indigestion remedies. They wanted to find out which one could neutralise most acid.

They crushed each tablet. Then they added them to beakers containing water and universal indicator solution.

Finally they added acid to each one to see how much the tablets could neutralise.

Here are their results:

Indigestion remedy	Volume of acid added (cm^3)
Fizzo	12
Rumtum	8
Coolers	7

a What type of substance do all the indigestion remedies contain? (1)

b Which one neutralised most acid? (1)

c Why did Nina and Alex use acid from the same bottle for each test? (1)

d How could they make their results more reliable? (1)

Key words

Unscramble these:

icad
catindiro
akilla
rcoorvies

7F Simple chemical reactions

Coming soon

Concrete setting is an example of a slow chemical reaction

Gunpowder exploding is an example of a fast chemical reaction

What's it all about?

In Science we have some changes that we call **chemical reactions.** These reactions are not always fast, like the ones you see in fireworks.

Our lives depend upon the chemical reactions inside our bodies. These reactions make new materials and also transfer energy to our cells.

In this unit you will learn more about chemical reactions and try some out for yourself.

What do you remember?

You already know about:

- some different gases
- changes in which new materials are formed and which, sometimes, cannot be easily reversed
- the pH scale as a measure of how strongly acidic or alkaline a solution is

1. Which one of these substances is a gas at room temperature?

   sulphur hydrogen
   sodium hydroxide oil

2. Which of these changes forms a new substance?

   ice melting alcohol boiling
   paper burning sugar dissolving

3. Which of these pH numbers shows that a solution is alkaline?

   1 4 7 11

Ideas about chemical reactions and substances

Chemical reactions in the kitchen? Surely we only do those in a science lab!
This must be a chemical reaction because the water has turned to ice.
This boiling water is giving off a gas... it must be reacting with the air
The pH of this solution is 14. It is a high number so it must be very acidic.
How can gases really be there... They don't weigh anything, do they?

QUESTIONS

Look at the cartoons above:
Discuss these questions with your partner.

a) Why are Pete, Reese, Molly and Mike wrong when they talk about:
  i) water freezing?
  ii) making toast?
  iii) water boiling?

b) What do you think a pH number of 14 tells us about a solution?

c) Do you believe gases really exist? Explain why.

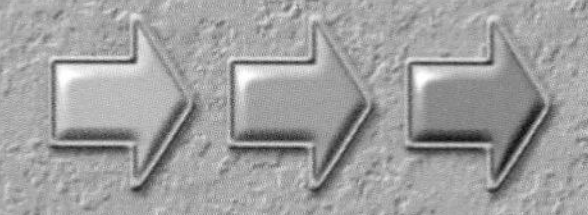

7F1 Observing chemical reactions

LEARN ABOUT
- making and interpreting observations
- recognising chemical reactions

What are reactions?

Lots of people enjoy fireworks on bonfire night. We can thank **chemical reactions** for all those bangs and flashes.

Remember that chemical reactions form new substances. They also involve energy changes. You often find energy given out as heat, and sometimes as light or sound.

Think of the sparklers you can buy.

Q1 What must you do to start off the chemical reaction in a sparkler?

Q2 Once the sparkler has finished, can you use it again? Why not?

The sparks from sparklers are tiny pieces of burning iron.

Heating iron wool

SAFETY

- Hold some loosely packed iron wool in a pair of tongs.
- Put the end of the iron wool into a Bunsen flame until it **ignites**.
- Take the iron wool out of the flame and watch the reaction.

SAFETY: Make sure you hold the burning iron wool over a heatproof mat.

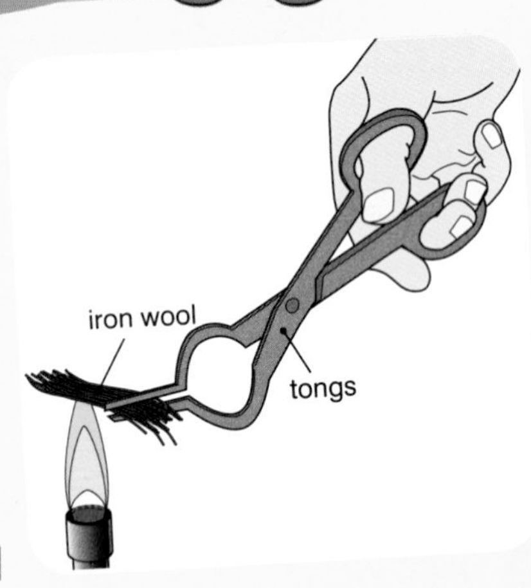

How can you tell that this chemical reaction gives out energy?

The new substance made in the reaction is called iron oxide. It is difficult to see because it forms as tiny bits of solid.

You can learn more about burning reactions on p.102.

Reactants and products

The new substances made in chemical reactions are called **products**.

The substances we start with are called **reactants**.

So in a chemical reaction:

reactants → products

Looking at reactions

Mix the following pairs of substances together. Then record all your observations in the table.

- plaster of Paris and water (in a yoghurt pot and stir with an ice-lolly stick)
- lemon juice and bicarbonate of soda (in a beaker)
- baking powder and water (in a beaker)

Reaction between	What happens? What do you see? How does the container feel?
plaster of Paris and water	
lemon juice and bicarbonate of soda	
baking powder and water	

a) Why was it easy to see that a new substance was made in the last two reactions?
b) Which reactions gave out heat and which felt cool?
c) Which reaction was slowest?

Can you help, I think this reaction has got out of hand!

SUMMARY QUESTIONS

1 ☆ Finish this sentence:

In a chemical reaction new ....

2 ☆ **a)** What do we call the substances that we start with before a chemical reaction takes place?
b) What do we call the substances formed in a chemical reaction?

Key words

ignite
product
chemical reaction
reactant

Acids and metals

LEARN ABOUT

- testing for hydrogen gas
- acids corroding metals
- recording observations

Corroding metals

In Unit 7E we saw how some acids are corrosive. But even very dilute solutions of acids can corrode materials, if given enough time.

You will have heard of **acid rain**. It can attack many metals and other building materials.

You can see how metals react with acids in the experiments below.

Zinc metal reacts with dilute acid

Magnesium plus acid

- Put about a 1 cm depth of dilute hydrochloric acid in a small beaker.
- Hold the end of a strip of magnesium ribbon in a pair of tongs.
- Dip the other end of the ribbon into the acid for a few seconds.

  a) What happens to the size of the magnesium ribbon? The acid **corrodes** the magnesium metal.

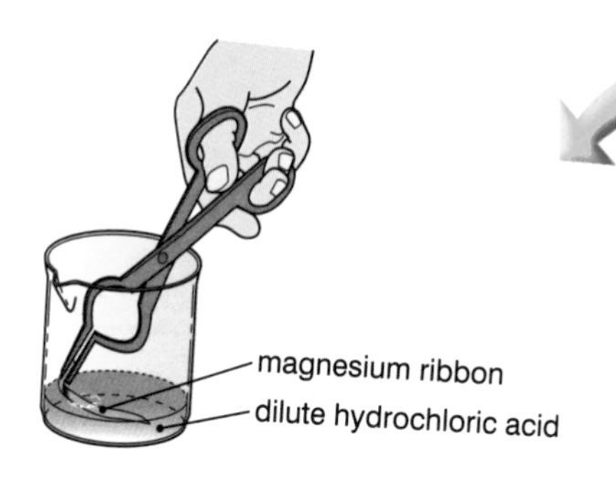

- Now put 5 cm^3 of dilute hydrochloric acid in a test tube.
- Take its temperature with a thermometer.
- Add a 2 cm strip of magnesium ribbon to the acid.

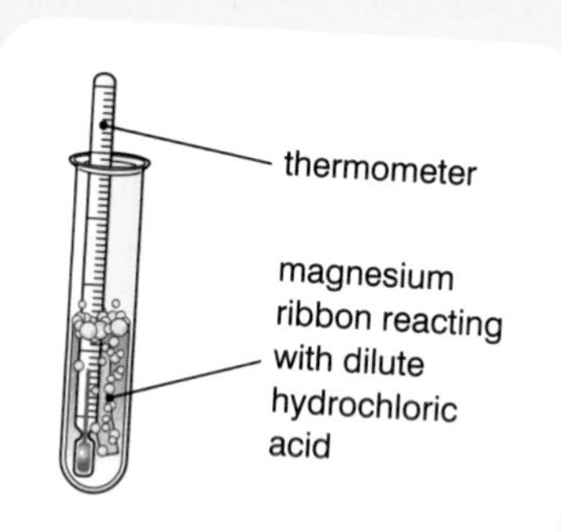

- Record the highest temperature of the reacting mixture.

  b) How can you tell that a chemical reaction takes place? Give two observations in your answer.

The gas we get when magnesium reacts with hydrochloric acid is called **hydrogen**.

Testing for hydrogen gas

- Repeat the reaction between magnesium and dilute hydrochloric acid. Use the apparatus shown in the first picture.
- Test the gas collected in the boiling tube as shown in the next picture. What happens when the hydrogen is tested with the lighted splint?

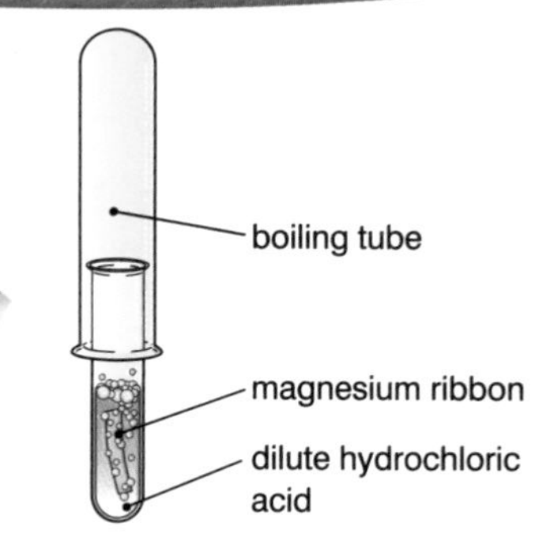

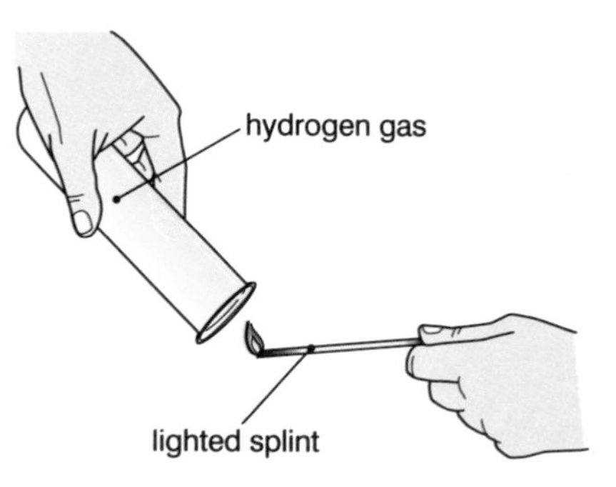

It's important to remember the test for hydrogen gas:

Test: hydrogen gas burns with a squeaky pop when we apply a lighted splint.

Investigating metals and acids

In this experiment you can test a range of metals with different acids. Look out for any patterns you can find in the reactions. Your teacher will give you a table to fill in your results.

- You will have the following metals:
  zinc copper iron magnesium
- You will also have these dilute acids:
  hydrochloric acid **sulphuric acid** **nitric acid**
- Now try out your tests.
  a) What usually happens when a metal is added to an acid? We call this a **generalisation**.
  b) Which metal does not fit into the general pattern?

SUMMARY QUESTION

1 ☆ Copy and complete the sentences, using the words from the list below.

hydrogen pop lighted acid

Many metals react with . . . to produce . . . gas.
We can test this gas with a . . . splint. The gas burns with a squeaky . . . .

Key words

acid rain
corrode
generalisation
hydrogen
nitric acid
sulphuric acid

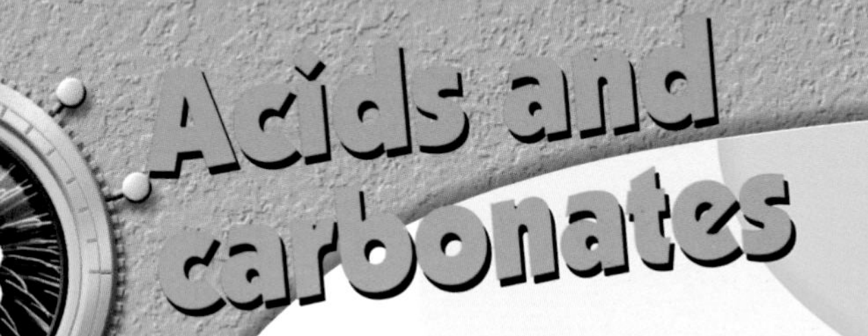

7F3 Acids and carbonates

Learn about

- testing for carbon dioxide gas
- uses of carbon dioxide

Carbon dioxide

Do you know which gas puts that 'fizz' in a fizzy drink? Fizzy drinks are also known as 'carbonated' drinks.

The name comes from the **carbon dioxide** gas dissolved in the drink.

Making and testing for carbon dioxide

- Set up the apparatus as shown in the picture and diagram.

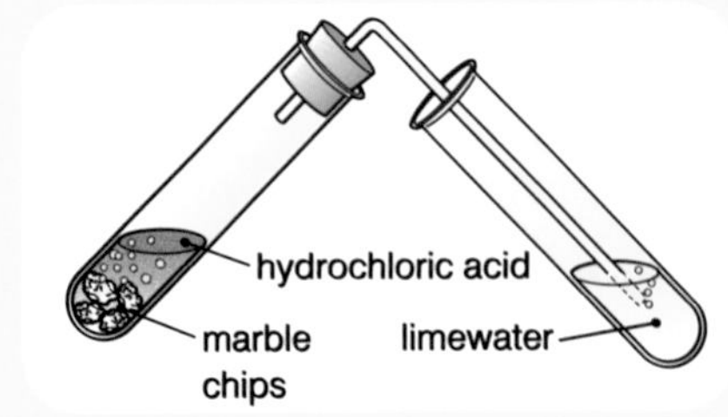

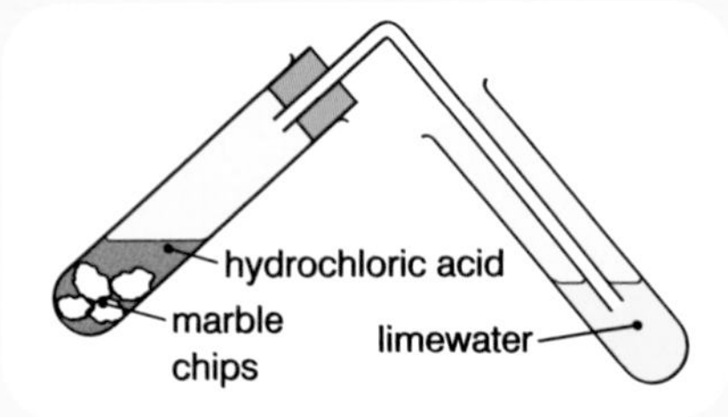

a) How can you tell that the marble chip and acid react together?
b) How can you tell when the reaction has finished?
c) What happens to the **limewater**?

- Now collect some carbon dioxide gas in a jar, or watch your teacher do the experiment below.

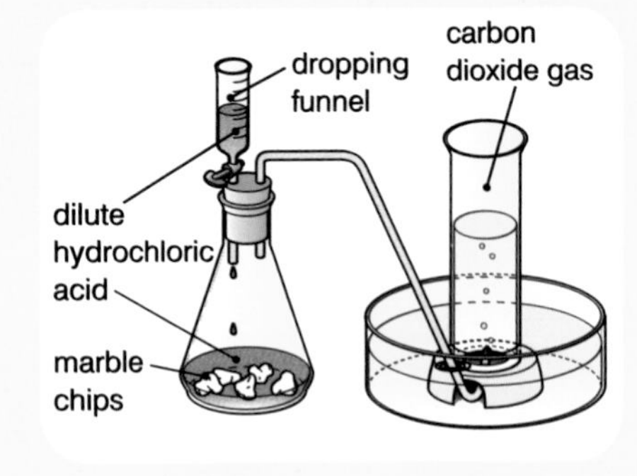

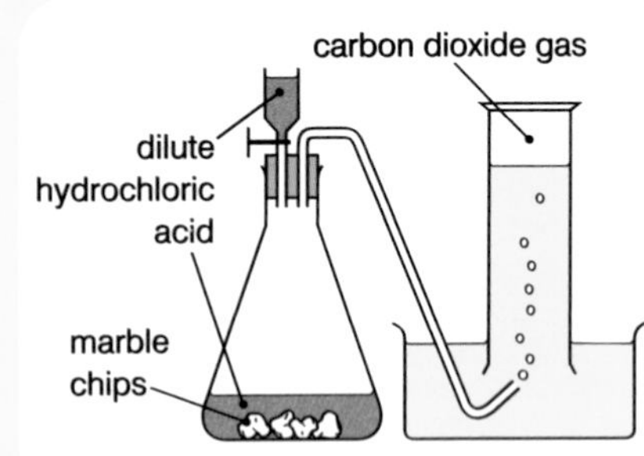

- Try 'pouring' a gas jar of carbon dioxide on to a burning night-light.

d) What does this test tell you about carbon dioxide?

You will need to remember the test for carbon dioxide.

Test: carbon dioxide turns limewater milky/cloudy.

Gruesome science

In the 1990s, almost two thousand people were found dead in the villages on the shores of Lake Nyos in Cameroon, Africa. Scientists eventually worked out that a cloud of carbon dioxide gas had killed them. The gas came from the lake after it was stirred up by a land-slide.

Reacting carbonates

Marble chips are made up mainly of calcium carbonate. There are also lots of other substances called **carbonates**.

Examples include copper carbonate, zinc carbonate and iron carbonate.

The question is:
Do all carbonates react with any acid to produce carbon dioxide gas?

Investigating carbonates and acids

- You will be given a range of different carbonates and different acids to test. Your teacher will give you a sheet to help you plan your tests. Let your teacher check your plan before you start.
- Think about these questions as you observe each carbonate react with an acid:
  a) Is a gas given off? If so, which gas is it?
  b) How can you test the gas?
- After your investigation, answer these questions:
  c) In what ways are the reactions similar?
  d) How do they differ?

Uses of carbon dioxide

Carbon dioxide gas is always given off when a carbonate reacts with an acid.

We use carbon dioxide in **fire extinguishers**. It is useful on electrical fires and many chemical fires. Water on these fires might make matters worse. Sometimes the carbon dioxide is trapped in foam so that it stays on the fire.

Q1 Why is it a good idea to use foam on an aeroplane fire?

SUMMARY QUESTION

1 ☆ Copy and complete the sentences, using words from the list below.

milky carbon limewater acid

All carbonates react with . . . to produce . . . dioxide gas.
We can test this gas by bubbling it into . . . which turns . . . .

Key words
carbon dioxide
carbonates
fire extinguisher
limewater

Combustion

Learn about

- what is needed for things to burn
- the new substances formed when things burn
- writing word equations

The fire triangle

In 7F3 we saw how carbon dioxide can put out fires.

A fire needs three things to keep it burning. We can show these in the **fire triangle**. If you remove one thing from the fire triangle, the fire goes out.

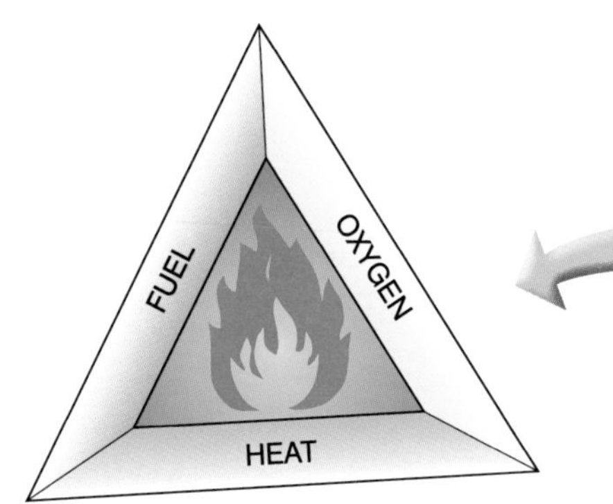

All three things in the triangle are needed for a fire

Carbon dioxide from a fire extinguisher stops oxygen gas getting to the burning fuel. Without **oxygen**, the fire goes out.

Research into fire fighting

- Use secondary sources to find information on how to extinguish different types of fire.
- You can use reference books, videos, CD-ROMs or the Internet.
- Working as a group produce a script for a TV advert on either 'Fire fighting' or 'Preventing fires'.
- Act out your advert for the rest of your class.
- Make sure that you explain clearly the science involved.

Gruesome science

In 1967, the three crew of the Apollo 1 spacecraft were killed when practising for their flight. They were in an atmosphere of pure oxygen. Then a small electrical fault produced a spark. The resulting flash fire killed them instantly.

Combustion reaction

Burning is a chemical reaction. The substance that burns reacts with oxygen. In this chemical reaction we get new substances called **oxides** formed. Energy is also given out during the reaction.

Burning is also known as **combustion**. A fuel releases energy as it reacts with oxygen in a combustion reaction.

You can try out a combustion reaction in the next experiment. Read through all the steps before starting.

Burning magnesium ribbon

- Hold a small piece of **magnesium** ribbon at one end with a pair of tongs.
- Ignite the other end in a Bunsen flame. Do not look directly at the burning magnesium.
- As soon as the reaction starts, take the magnesium out of the Bunsen flame.
- Hold the burning magnesium above your heatproof mat. Make sure it is well away from the Bunsen tubing.

a) Describe what happens in the reaction.
b) How does the product differ from the magnesium you started with?
c) Which gas in the air did magnesium react with?

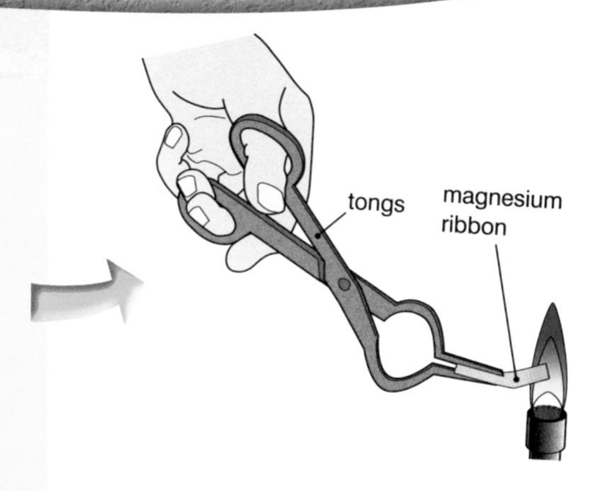

Word equations

We can show the substances involved in a chemical reaction in a **word equation**. A word equation tells us the reactants we start with and the products they make. We can show the combustion of magnesium as:

magnesium + oxygen → magnesium oxide

Q1 Name the reactants in the combustion of magnesium.

Q2 Name the product in the combustion of magnesium.

Magnesium powder is used in fireworks

SUMMARY QUESTION

1 ☆ Copy and complete the sentences, using the words in the list below.

oxygen combustion oxides

The chemical name for burning is . . .
When things burn, they react with . . . gas in the air.
The products formed are called . . .

Key words

combustion
fire triangle
magnesium
magnesium oxide
word equation
oxides
oxygen

Burning fuels

LEARN ABOUT
- the products formed when we burn fuels
- investigating combustion

Fires and fuels

Have you ever seen a car fire? Sometimes cars catch fire because of faults or after accidents. Sometimes stolen cars are set alight deliberately. Look at the fire in the photo.

Fuels are substances that release energy when we burn them. We can then use the energy to do some useful job, such as heating our homes.

Most of the fuels we use are **fossil fuels** (or we get them from fossil fuels). Examples of fossil fuels are **coal**, **oil** and **natural gas**.

Q1 List three fuels we can use to heat our homes.

Q2 Fossil fuels contain lots of carbon. What might be formed when we burn a fossil fuel?

Burning methane

- Watch your teacher carry out the experiment shown in the picture.

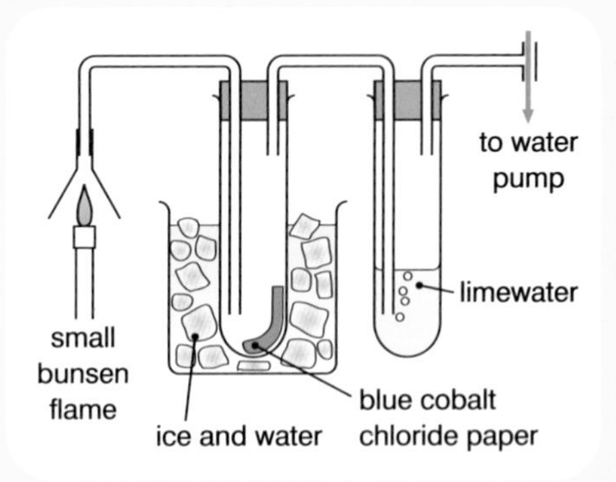

Test: water turns blue cobalt chloride paper pink (or white anhydrous copper sulphate blue).

a) What do you see happen in the experiment?

b) Why do we put ice around the first tube?

c) What products do we get when methane burns?

The combustion of wax is a very useful reaction sometimes.

Products of combustion

When a fuel containing carbon burns, the carbon turns to carbon dioxide. (This happens if there is plenty of **oxygen** around for the fuel to react with.)

As well as carbon, fossil fuels also contain hydrogen. This reacts with oxygen in the air to form water (hydrogen oxide).

Bunsen burners usually use natural gas as their fuel. Natural gas is made up mainly of a gas called **methane**. Methane contains carbon and hydrogen. So when methane burns it makes carbon dioxide and water.

What happens to the air?

We have seen that substances react with the oxygen in air when they burn.

- Watch the experiment shown in the picture.

Record your observations.

Try to explain your observations.

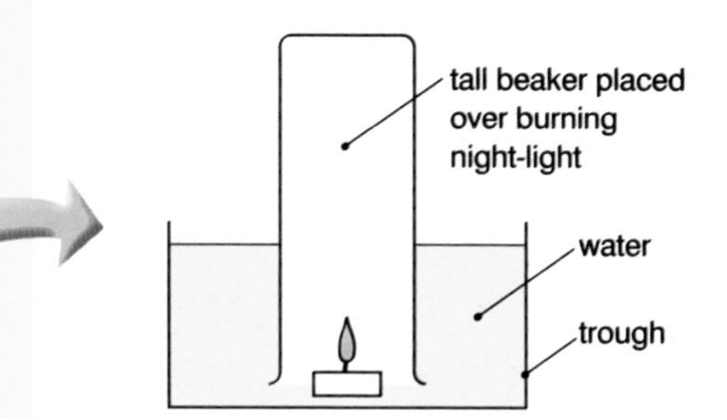

Investigating burning

- Your task is to find out: *How does the volume of air affect the time a night-light burns?*

You can use a variety of different-sized beakers in your tests.

Your teacher will give you a sheet to help you plan and write up your investigation.

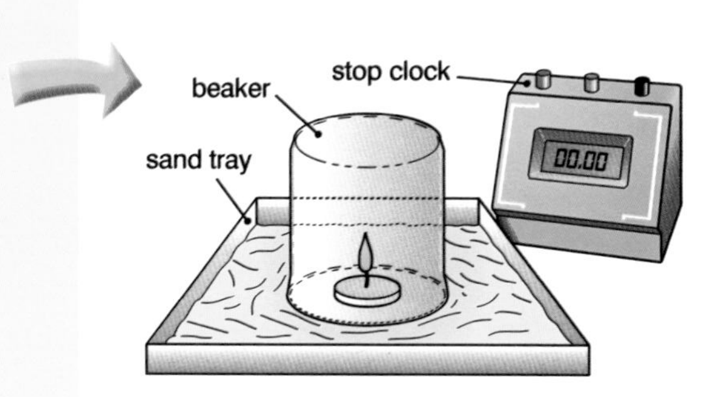

SUMMARY QUESTION

1 ☆ Copy and complete the sentences, using the words in the list below.

water   carbon dioxide   energy   oxygen
combustion   carbon

Fossil fuels contain . . . and hydrogen. When the fuels burn they react with . . . gas releasing . . . . The products of the . . . reaction (in plenty of air) are . . . and . . . .

Key words

coal
combustion
fossil fuel
methane
natural gas
oil
oxygen

7F
Read all about it!
IDEAS AND EVIDENCE

The Hindenburg disaster

Have you seen the airships that sometimes fly over big sporting events? The amazing overhead shots you see on TV are taken from airships. Modern airships are filled with a gas called helium. They are a very safe way to travel. However, this was not true of the giant airships built in the 1930s.

GOODYEAR

A modern airship, filled with helium gas

Airships are really huge balloons fitted with engines that drive propellers. The Germans were the masters of airship technology. They used them in World War 1 to drop bombs on London. After the war, the Germans went on to make airships for passenger flights. Most of the airships were made by the Zeppelin Company. Their biggest airship was the *Graf Zeppelin*. It was like a great flying hotel. It flew right around the world – although it did take 3 weeks!

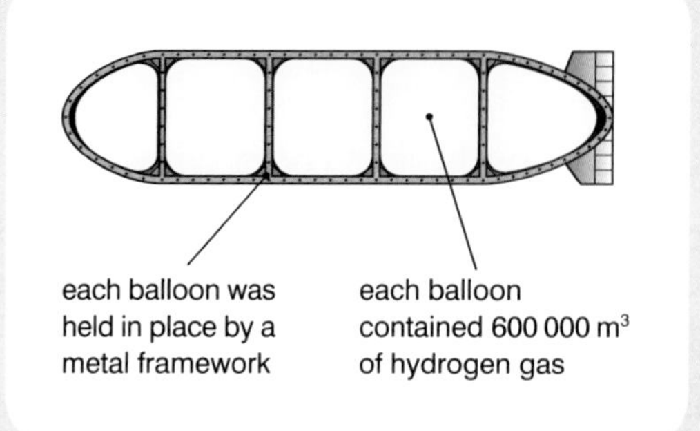
each balloon was held in place by a metal framework
each balloon contained 600 000 m³ of hydrogen gas

The airships were filled with the lightest of all gases, hydrogen (not helium as nowadays). The British built two large passenger airships too – called the *R100* and the *R101*. But the *R101* crashed in France on 4 October 1930. It was on its way to India and 48 people died. After that, travel by airship was left to the Germans.

However, disaster struck in 1937. Their newest and best airship, called the *Hindenburg*, crashed as it landed in New Jersey, USA. It was a stormy day, but crowds had still gathered to see the great airship arrive. They couldn't believe their eyes when it burst into flames on landing. After about 30 seconds, only its skeleton was left.

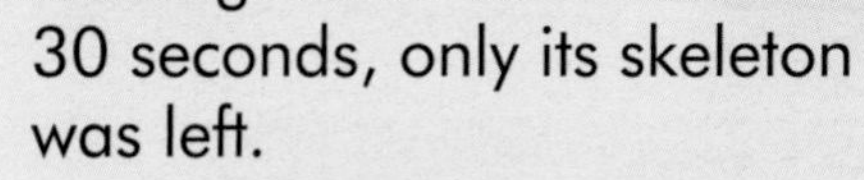

The hydrogen gas burned fiercely. It reacted with the oxygen in the air. Remember:

hydrogen + oxygen → water
(*hydrogen oxide*)

36 people died at the airfield. Remarkably 62 others survived, although some died later of their injuries.

Hydrogen was just too dangerous to use in airships. The *Hindenburg* was the last of the giant airships.

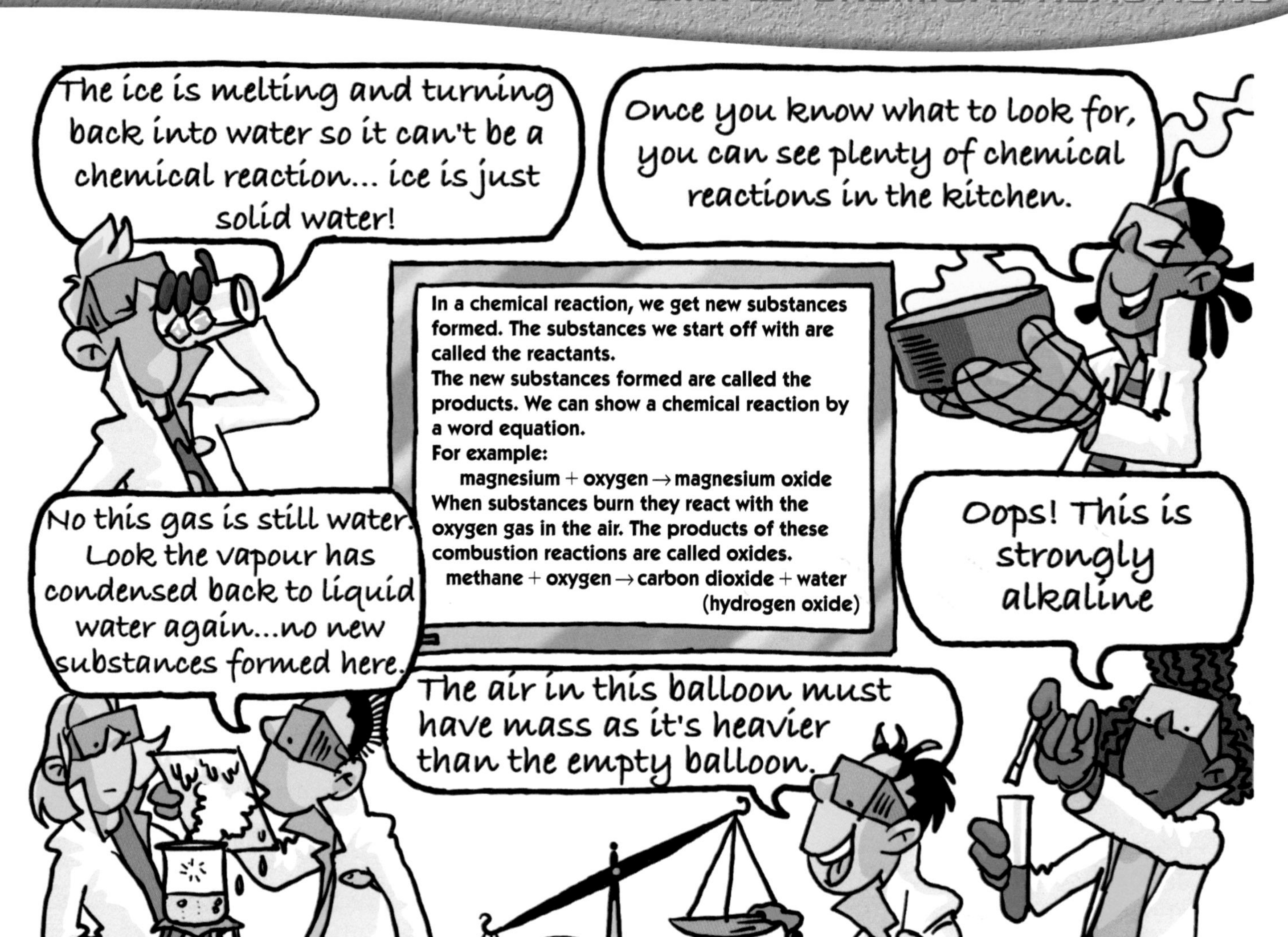

In a chemical reaction, we get new substances formed. The substances we start off with are called the reactants.
The new substances formed are called the products. We can show a chemical reaction by a word equation.
For example:

magnesium + oxygen → magnesium oxide

When substances burn they react with the oxygen gas in the air. The products of these combustion reactions are called oxides.

methane + oxygen → carbon dioxide + water (hydrogen oxide)

There are two types of change that chemists investigate:

- **chemical changes** (reactions)
- **physical changes** (such as melting and boiling).

The main difference is that new substances are formed in chemical changes, but no new substances are formed in a physical change.

Key words

combustion
oxygen
products
reactants
reaction

UNIT REVIEW

REVIEW QUESTIONS

Applying and understanding concepts

1 Copy and complete the sentences, using words from the list below.

pop combustion oxides chemical limewater oxygen substances hydrogen oxide carbon products word

We get new ... formed in a ... reaction. We start off with the reactants and end up with the ... of the reaction.

A ... equation describes a reaction. For example:

zinc + oxygen → zinc ...

If a metal reacts with an acid we get ... gas given off. This gas will burn with a ... when we apply a lighted splint to it.

Acids also react with carbonates, giving off ... dioxide gas. This gas turns ... milky.

The chemical name for burning is .... When a fuel burns it reacts with the ... gas in the air and forms ....

2 Copy and complete the following word equations:

- **a** copper + ... → copper oxide
- **b** sodium + oxygen → ...
- **c** ... + oxygen → water (... oxide)
- **d** carbon + ... → ... dioxide

Ways with words

3 Write a short poem about burning. Make sure you use your knowledge of the fire triangle in your poem.

FUEL
OXYGEN
HEAT

4 Match the words **a** to **d** with the statements **1** to **4**:

- **a** fuel
- **b** oxygen
- **c** limewater
- **d** ignite

1. a solution that we use to test for carbon dioxide
2. to set a substance alight
3. a substance that burns to give us useful energy
4. the reactive gas in the air

Making more of maths

5 Crude oil is a source of many different fuels. The table shows different fuels we get from a sample of crude oil.

Draw a bar chart to display the data.

Fuel	Approximate percentage by mass
Gas	3
Petrol	25
Kerosene	10
Diesel and gas oil	20
Fuel oil and residue	42

Extension question

6 Find out about the chemical reaction used by brewers to make alcohol and by bakers to make bread rise. Draw a poster about the reaction to help shoppers understand that chemical reactions play an important part in everyday life.

SAT-STYLE QUESTIONS

1 Here is a key you can use to identify gases.

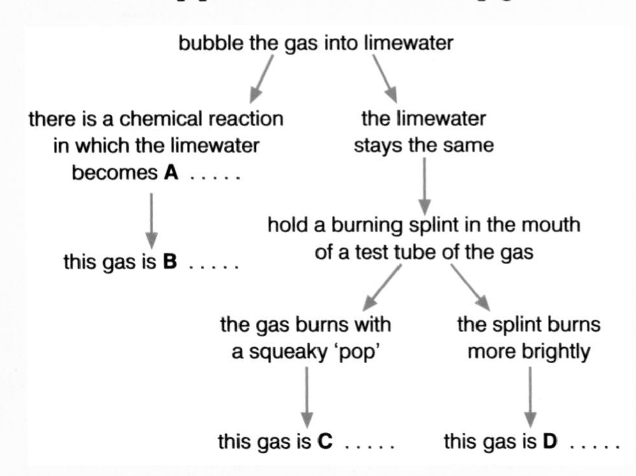
bubble the gas into limewater
there is a chemical reaction in which the limewater becomes **A** .....
the limewater stays the same
this gas is **B** .....
hold a burning splint in the mouth of a test tube of the gas
the gas burns with a squeaky 'pop'
the splint burns more brightly
this gas is **C** .....
this gas is **D** .....

a What are the missing words A to D? Choose from this list:

hydrogen oxygen
carbon dioxide milky (4)

b i) What would you see if you burned a piece of magnesium ribbon in gas D? (1)

ii) What safety precautions would you (or your teacher) take when doing this experiment? (2)

iii) Finish the word equation for the reaction in part (i).

magnesium + ... → ... ... (1)

2 Simon timed how long a candle burned under different beakers.

He measured the volume of each beaker. He did each test once.

Here is a graph of his results:

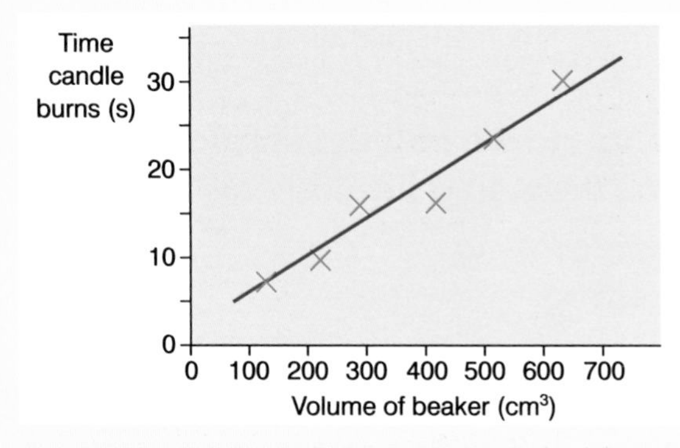
Time candle burns (s)
0
10
20
30
0 100 200 300 400 500 600 700
Volume of beaker (cm³)

a How long would a candle burn under a beaker with a volume of 250 cm^3? (1)

b How could Simon get more reliable results? (1)

3 A candle is burned under a gas jar in a sand tray.

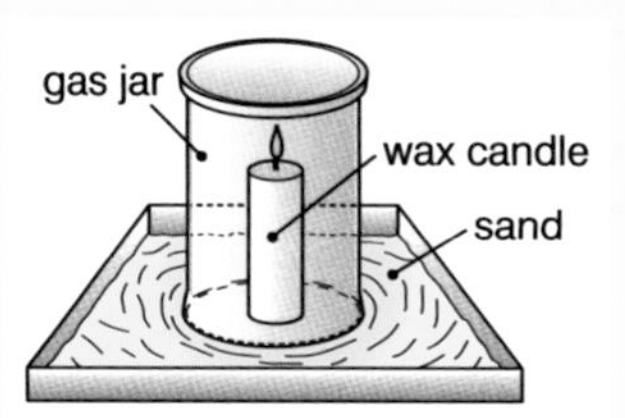
gas jar
wax candle
sand

a When the candle burns a chemical reaction takes place. Which gas is the candle wax reacting with? (1)

b As the candle burns, wax gets used up. Name two other ways in which we can tell that a chemical reaction takes place. (2)

c Explain what you see happen in the experiment above. (1)

d What safety precaution has been taken in the experiment? Why? (1)

4 Assam and Mia did an experiment to find out what products form when wax burns.

They set up the apparatus below:

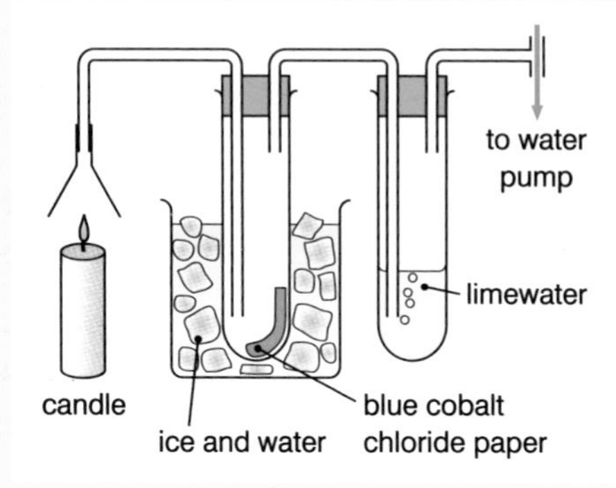
to water pump
limewater
candle
ice and water
blue cobalt chloride paper

a What happens to the limewater? (1)

b What does this show? (1)

c Why is the other tube surrounded by ice? (1)

d What to we call the type of reaction that takes place between wax and oxygen? (1)

Key words

Unscramble these:
engoyx
trodcups
acorinte

7G

The particle model

Coming soon

What's it all about?

Sorting things into groups is one way we have of making sense of the world. For example, we have solids, liquids and gases.

But people also want to *explain* how materials behave. That's where the theory of particles comes in.

In this unit you will be looking at evidence from experiments and thinking how to explain your observations.

Point out the solids, liquids and gases in this volcanic eruption

What do you remember?

You already know:

- the properties of solids, liquids and gases
- that the same material can exist as a solid, a liquid and a gas
- that melting and freezing are opposites
- about evaporation and condensation

1. What do we call the change when a liquid turns to a solid?

   boiling condensing
   freezing melting

2. What do we call the change when a gas turns to a liquid?

   evaporating boiling
   melting condensing

3. Which of these is a property of a gas?

   It has a fixed shape.
   It cannot be compressed.
   It has a very high density.
   It fills the shape of its container.

Ideas about solids, liquids and gases

He
I thought all gases floated as they don't have any mass, do they?
CO_2
Don't try telling me that everything is made of particles if you can't show me a particle right now!

The particles that make up ice, water and steam must be different. I bet steam has really light particles but ice has much heavier particles.
I think if you heat something up, its particles will blow up like mini-balloons.
Is that why things expand when they get hot?

QUESTIONS

Look at the cartoons above:
Discuss these questions with your partner.

a) Why do you think some gases float and others don't?

b) Do you believe that everything is made up of particles? Do you believe in anything you've never seen with your own eyes? Try explaining why a piece of elastic stretches when you pull it.

c) What do you think of Benson's idea about the particles in ice, water and steam?

d) Does Pete's and Molly's idea about hot materials make sense to you? Why?

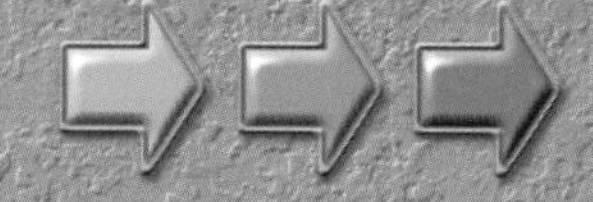

7G1 Evidence from experiments

LEARN ABOUT
- classifying materials as solid, liquid or gas
- interpreting and explaining results
- how theories can be based on experimental data

No, don't shout... she's only experimenting!

Solids, liquids and gases

You start experimenting from a very early age. A toddler soon enjoys shaking her drinking cup upside down. It makes a nice puddle in the tray of her highchair. This is great for splashing her hands in – and much more fun than drinking!

In this way we find out that liquids spread out easily, can flow and don't have a fixed shape.

This table gives a summary of the properties of solids, liquids and gases.

	Does it have its own fixed shape?	Is it easy to compress?	Does it spread out or flow easily?
Solid	yes	no	no
Liquid	no	no	yes
Gas	no	yes	yes

Q1 Which are solids, liquids or gases at normal room temperature?

petrol oxygen nitrogen concrete cooking oil iron carbon dioxide Perspex vinegar

Discussing ideas

Have you ever tried to run through water in a swimming pool? It's hard work!

- But why is it so much harder to get through water than it is to move through air?
- Talk about your ideas in pairs. After a few minutes share your ideas with a neighbouring pair. Do you all agree why it is easier to move through air than water?
- Draw a large diagram to display your ideas to the rest of the class.
- Discuss the ideas from other groups. Compare them with your own.

Investigating solids, liquids and gases

SAFETY

- Try to push in the plunger of the three sealed syringes.
  - **a)** What do you feel?
  - **b)** Explain your observations.

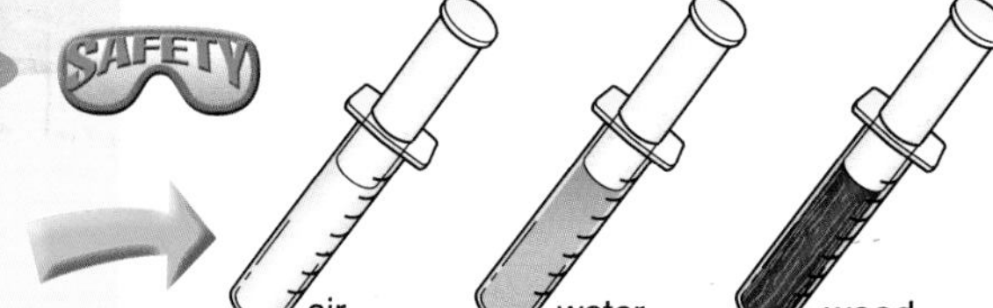

- Try the 'bar and gauge' experiment or the 'ball and ring' experiment.
  - **c)** Explain what happens.

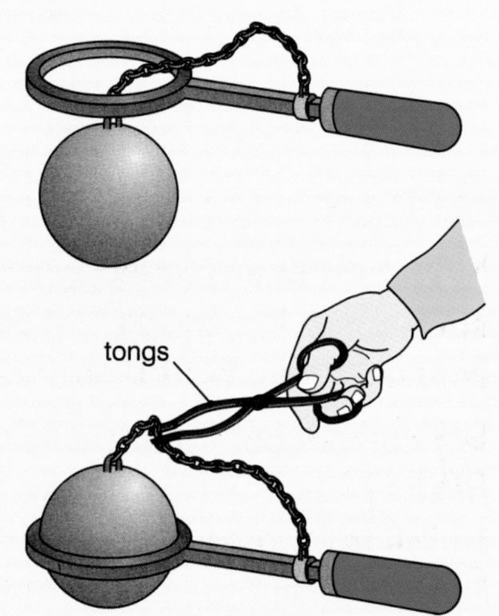

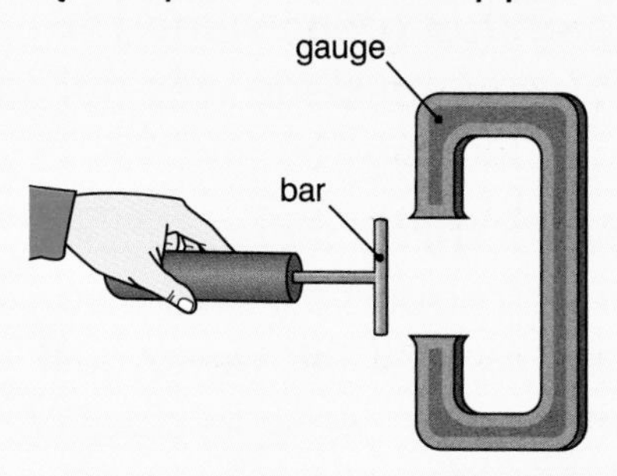

- Use tweezers to place one coloured crystal at the bottom of a beaker of cold water. Now place one at the bottom of a beaker of hot water.
  - **d)** Explain your observations.
- Compare the masses of blocks of materials that are the same size.
  - **e)** Put your results in a table.
  - **f)** Explain your data.

Discuss your explanations together as a whole class.

The Forth Rail Bridge is made of iron. In the summer it is about 50 cm longer than it is in the winter.

SUMMARY QUESTION

1 ☆ Copy and complete the sentences, using words from the list below.

spread shape stay shape compressed

Solids and liquids cannot be . . . easily, unlike a gas.

A solid has a fixed . . . , but liquids and gases take up the . . . of their container.

A gas will . . . out in all directions, but solids . . . in one position.

Key words

compress
gas
liquid
solid

Classifying materials

LEARN ABOUT

- classifying materials as solid, liquid or gas
- some materials being difficult to classify
- how theories can be based on data from experiments

I told you it was a liquid... it flowed quite nicely on your face.

States of matter

Have you ever played with a tub of that green slime you can buy in toy shops? Would you say the slime was a thick (**viscous**) liquid or a soft solid?

We call solids, liquids and gases the **three states of matter**. In 7G1 we looked at their properties. We also classified materials as solid, liquid or gas. However, just like green slime, some things are not that easy to classify.

Classifying tricky materials

Copy and complete a table like this:

Material	I would classify this as a . . . (solid/liquid/gas)	because . . .

Materials to try classifying:

flour sponge hair gel hair mousse
whipped cream plastic film glue-stick
wallpaper paste hair spray toothpaste

Scientists use physical models like this to help explain the world

Models in science

Scientists use models to help them explain the way the world works.

A scientist's model can also be:

- a **theory**
- a mathematical equation
- a computer simulation.

The model should explain our observations. Then we can use the model to make **predictions**.

But is the model a useful one?

If predictions prove to be correct, the **model** is more likely to be accepted by other scientists.

Imagine that you have the most powerful microscope ever invented. It is so good that it can look right inside materials.

Q1 What do you think you would see?

Q2 How would your model explain the differences between solids, liquids and gases?

Get together with others in your class. Share your different theories. Discuss how people have used **evidence** to support their model.

ICT CHALLENGE

Devise a branching database that could be used to classify materials as solid, liquid or gas. You will need to think up questions that can be answered with a 'Yes' or 'No', leading to the word 'solid', 'liquid' or 'gas'.

Changing models

The Ancient Greeks were the first to suggest that everything is made of tiny particles. The Greeks were great thinkers about the world they lived in. However, they didn't bother much with experiments.

A Greek called Democritus put forward his model. It explained the way materials behave. The **particles** he talked about were so small that you couldn't see them.

Nowadays most people agree that the particle model is a good one. But the model has changed since Greek times.

SUMMARY QUESTION

1 ☆ Copy and complete the sentences, using words from the list below.

Greeks particles models

Scientists use . . . to help explain the way materials behave.

The Ancient . . . were the first to think that everything is made of tiny invisible . . . .

Key words

evidence
model
particles
prediction
theory
viscous

7G3 The particle theory

LEARN ABOUT

- using a model based on things we can't see directly
- explaining the differences between solids, liquids and gases

Can we see the particles?

Scientists in the 1800s found new evidence that particles exist. In 1805 John Dalton explained chemical reactions using ideas about particles.

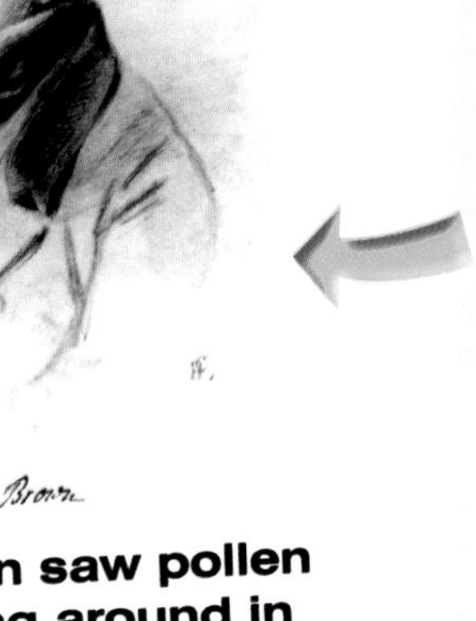

Robert Brown saw pollen grains jiggling around in water

About 20 years later Robert Brown actually observed the effects of particles. He looked at pollen grains under a microscope.

He noticed that pollen grains in water appeared to jiggle about constantly. Their movement was **random**. It was eventually explained using the **particle theory**.

This said that the pollen grains were being hit by the particles that make up water. We can't see these particles because they are too small.

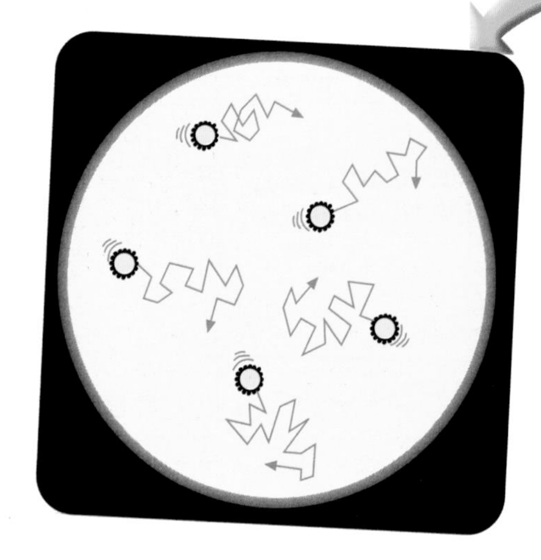

Models of solids, liquids and gases

Solid

We can think of the particles in a solid as being very close together. Each particle is touching its neighbouring particles. They are fixed in position. However, they are not still. They vibrate constantly.

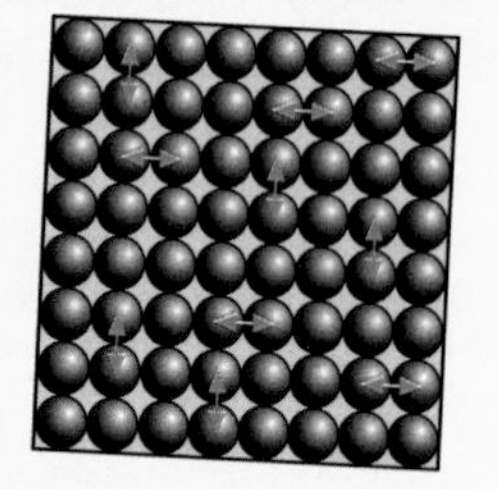

Q1 What do you think happens to the vibrations of the particles if you heat a solid?

Liquid

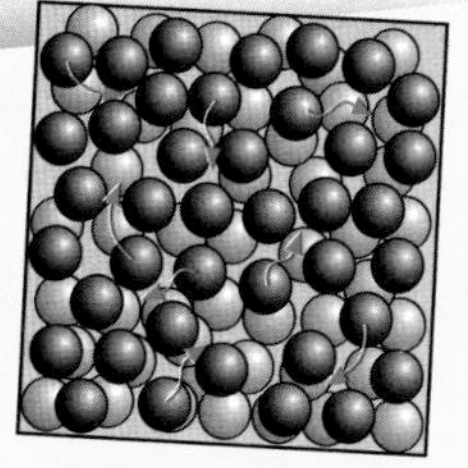

The particles in a liquid are still very close together. You can imagine the particles a bit like a bunch of grapes. They touch each other but there is no regular pattern. Unlike the particles in a solid, they are *not* fixed in position. They are free to slip and slide over and around each other.

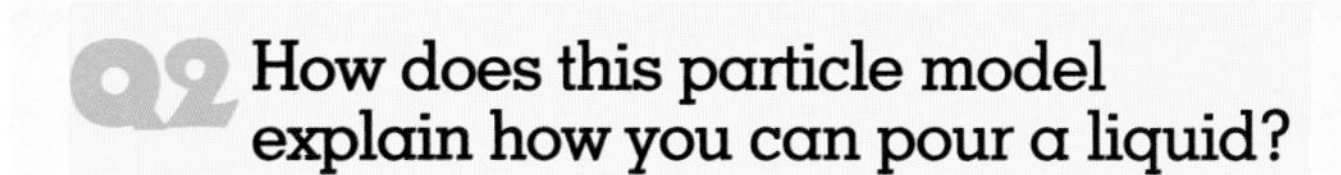

Q2 How does this particle model explain how you can pour a liquid?

Gas

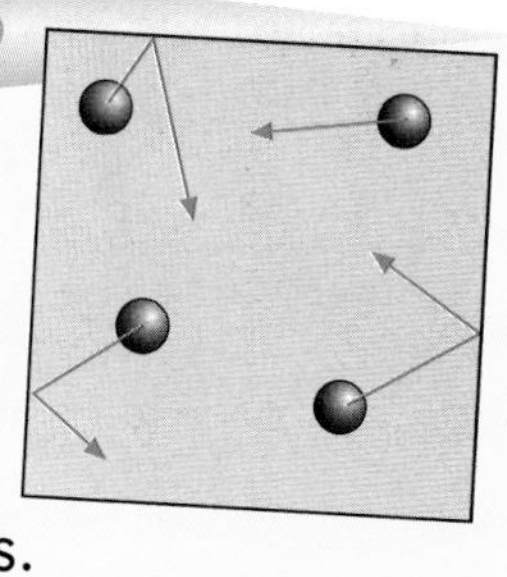

The particles in a gas are free to zoom around in any direction. On average, there is plenty of space between the particles. They often collide with other particles in the gas. They also bash into the walls of their container.

Q3 How does this model explain why it is easy to move through a gas?

Role play – Particular people!

- Imagine that the people in your class are particles. Use them to show the arrangement and movement of particles in a solid, liquid and gas.
- Think up a set of instructions for this modelling activity.

AMAZING SCIENCE!

The particles we are talking about in this unit are very, very small. If we could line up a couple of million average-sized particles, they wouldn't even measure 1 mm from one end to the other. That's why we can't see them!

SUMMARY QUESTION

1 ☆ Copy and complete the sentences, using the words in the list below.

close particles vibrate very close fixed walls zoom particles slide

The . . . in a solid are very . . . together.

They are . . . in position but they do . . . .

In liquids the particles are still . . . together, but can slip and . . . over each other.

In a gas the particles . . . around, knocking against other . . . and the . . . of their container.

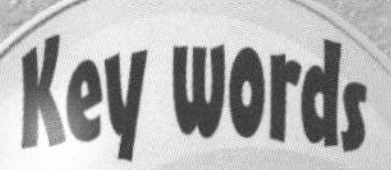

Key words

gas
liquid
random
solid
vibrate

Applying particle theory

LEARN ABOUT
- using particle theory to explain observations
- diffusion
- gas pressure

Can you believe it?

At first it might be hard to imagine a world made up of tiny particles. Particles that are too small to actually see. However, when you start using the particle theory, you'll find it is easier to believe. Particles help us explain our observations.

As we heat up a material, we transfer energy to its particles. This makes their movement more vigorous.

Q1 Use the particle theory to explain what happens when you heat butter gently in a pan.

Diffusion

Do you like the smell of freshly baked bread? You can thank a process called **diffusion** for spreading the smell. The particles from the bread travel through the air and into your nose.

Diffusion happens when substances mingle and pass through each other. It happens without us having to stir the substances up.

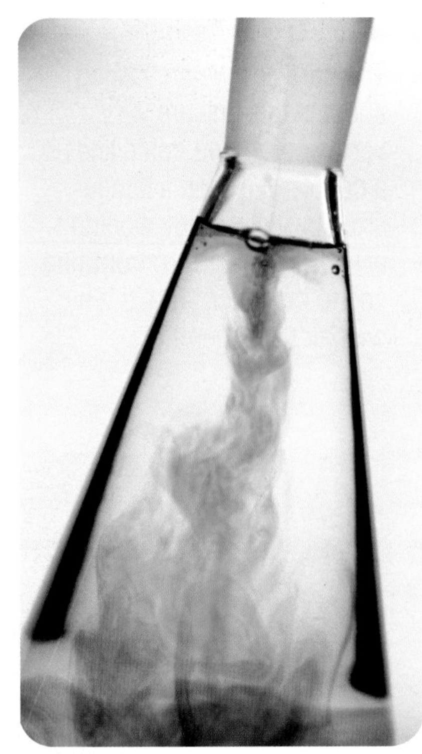

This green dye is diffusing through water

Looking at diffusion

Bromine is a dark brown liquid that evaporates easily. The gas it gives off is dark orange. The gas is much denser than air.

Use the particle theory to explain the following experiment.

- Your teacher will place a little bromine in the bottom of a gas jar. Then a second gas jar is placed on top of the first one.
  a) What do you think will happen?
  b) Why is the experiment done in a fume-cupboard?
  c) Record and explain your observations.

Gas pressure

Can you feel millions of gas particles bashing into you now? Of course, we don't feel these collisions, but they are happening.

As you know, the particles in a gas whizz around very quickly. They collide with each other and anything else they bump into. That includes the walls of their container.

Each collision will result in a tiny force being applied. This force causes **gas pressure**.

Gas pressure tells us how the force exerted by particles of gas is spread over an area.

Gruesome science

One of the most foul-smelling substances on Earth is ethyl mercaptan. Its smell has been described as a mixture of garlic, onions, rotting cabbage and sewer gas. Yeuk!

SUMMARY QUESTION

1 ☆ Copy and complete the sentences, using words from the list below.

pressure gases diffusion particles move force container stir

When substances mix by themselves, without us having to . . . them, we call it . . . . This happens because particles in liquids and . . . are free to . . . around randomly.

When the ... of a gas collide with the walls of their . . . they produce a . . . that causes gas . . . .

Key words

bromine
collision
diffusion
gas pressure

Otto von Guericke and his amazing scientific demonstrations

The name of Otto von Guericke will always be linked to the city of Magdeburg in Germany. Otto was born on 20 November 1602, but his wealthy family had lived there for three centuries. He went to university when he was 15 years old. Later he became mayor of the city for over 25 years.

Otto is famous for his demonstrations involving air pressure. After inventing the air pump in 1650, he learned how to create a vacuum. By pumping air out of a container, he could show the great force that air pressure can produce. And Otto really knew how to impress people. Look at the picture of one of his experiments!

One man, with the help of air pressure, proved stronger than ten men

In another experiment, Otto planned a tug-of-war between two teams of six horses. First he asked someone to make him a copper sphere in two halves. They fitted together perfectly. The local blacksmith and some helpers used a pump to suck the air out of the sphere. At his signal, the two teams of horses pulled and pulled. However, they could not separate the two halves of the sphere.

The crowd of curious on-lookers cheered. They were even more impressed when the horses stopped pulling. That's when Otto did his trick. He asked for silence as he released a valve. A hissing sound could be heard. As the air rushed back inside the sphere, it suddenly fell into two halves. The astounded crowd went wild.

Even Helium gas has mass... It's just very light. There's a lot of space between the particles of gas in both balloons but the actual particles in my balloon are lighter than yours.
He
CO_2

This particle model we've used really can explain a lot of stuff, you know.

- In a **solid**:
  - the **particles** are lined up next to each other
  - the particles are fixed in position, but they do vibrate
  - the particles vibrate more and more vigorously as the solid is heated.
- In a **liquid**, the particles are still very close together, but they can slip and slide over each other.
- In a **gas**:
  - the particles whizz around, and there is lots of space inside the gas
  - the particles collide with the walls of their container, and they produce a force that causes **gas pressure**.

It's just the arrangement and movement of the particles that change when a material melts or boils... The particles themselves don't change... That would only happen in a chemical reaction.

The particles vibrate more when they get hot so the gaps between them get a bit bigger...That's why a hot solid takes up more space than when it's cold.

DANGER! AVOID THESE COMMON ERRORS

We can compress gases a lot. But liquids are very difficult to compress. That's because the particles in a liquid are very close together.

The particles in a liquid and a gas can both move about randomly. The particles in a gas move a lot more quickly though.

When we heat a solid, its particles gain energy and vibrate more vigorously. In liquids and gases, heating makes the particles move around more quickly. This extra movement means that substances expand as they get hotter. The particles themselves do NOT expand.

Key words

gas
gas pressure
liquid
particles
solid

UNIT REVIEW

REVIEW QUESTIONS
Applying and understanding concepts

1 Copy and complete the sentences, using words from the list below. Use the summary on page 121 to help you.

vibrate pressure around walls quickly gas space solid

The particles in a . . . are fixed in position, but they do . . . .

In a liquid, the particles can move . . ., slipping and sliding over each other.

In a . . ., the particles move very. . ., and there is a lot of . . . between particles (on average). As they knock into the . . . of their container they produce a force that causes gas . . . .

2 On motorway bridges, the concrete sections are slotted together using expansion joints. You can see one in the photo.

In hot weather the road expands.

Explain why tar is used in the expansion joints.

3 Label your links in the concept map below:

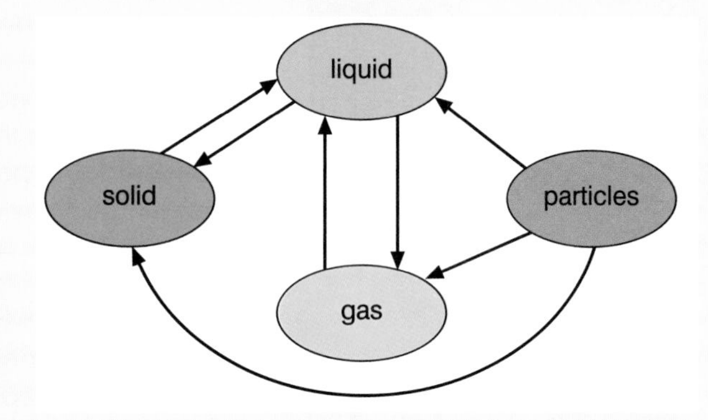

4 **a** How can you smell a fish and chip shop from across the road?

b Petrol is stored underground in large tanks under petrol stations. So why are there 'No smoking' signs in all petrol stations?

c What do we call it when two substances mix themselves without us having to stir them up?

5 A Year 7 class is going to model solids, liquids and gases using marbles and the lid of a shoe box.

Draw diagrams that show the arrangement of the marbles in the model of:

a a solid

b a liquid

c a gas

Ways with words

6 Write a short story or cartoon strip about a water particle that starts off in an ice cube inside a freezer. The next day it ends up in a drop of water on the inside of the kitchen window. Tell the tale in '24 hours in the wonderful world of a water particle'.

SAT-STYLE QUESTIONS

1 Solid, liquid and gas are called the **three states of matter**. The particles in a solid, liquid and gas are shown below. The arrows represent changes of state.

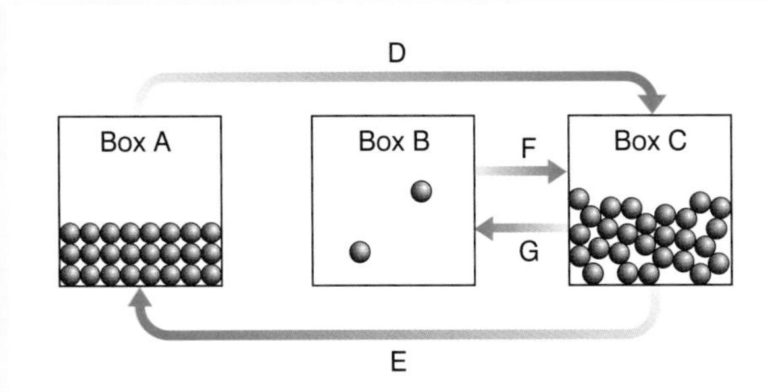

a Which box contains
 i) a solid
 ii) a liquid
 iii) a gas (1)

b Which state of matter is most easily compressed? (1)

c i) Identify the changes of state labelled D, E, F and G. (4)
 ii) Which changes of state require cooling down to take place? (2)

2 Sima wanted to see how easily sulphur powder melts. Sulphur burns in air to form toxic sulphur dioxide gas.

a Give two safety precautions that Sima should take in her experiment. (2)

b The sulphur melted quite easily. Sima stopped heating it and let it cool down to room temperature.
 What happens to the sulphur as it cools down? (1)

c If sulphur burns in a lab, you can soon smell the gas in all parts of the room.
 What do we call the process by which one gas mixes with another? (1)

d Sulphur melts at 119°C. Sima thought it might be safer to heat the sulphur gently in a test tube placed in a water bath.
 Why wouldn't her experiment work? (1)

3 Jason and Sharni are testing four metals (A, B, C and D). They set up the apparatus below and time how long it takes for each drawing pin to drop off the different metals:

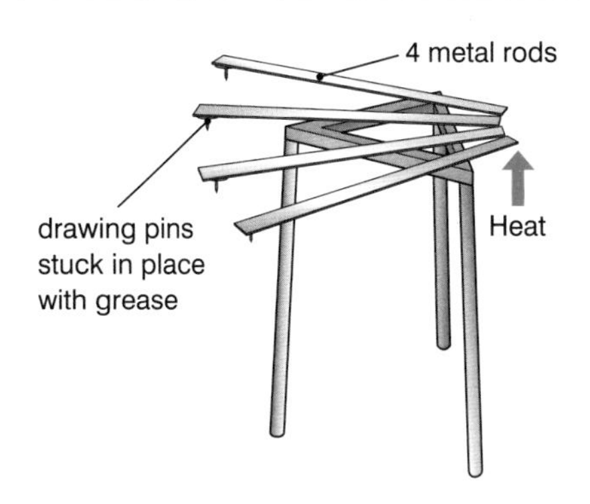

Here are their results.

Metal	Time for drawing pin to fall (s)
A	550
B	470
C	360
D	545

a Finish off the question that Jason and Sharni were investigating?
 Which metal is the best . . .? (1)

b How can they make their results more reliable? For which two metals is this particularly important?
 Why? (2)

c Give two ways in which Sharni and Jason try to make their investigation a fair test? (2)

d Before their tests, Jason and Sarni felt the metals. Sharni predicted 'I think metal C will be best because it feels coldest.'
 Do their results support her prediction? Explain your answer. (1)

Key words

Unscramble these:

ags
dilos
diquil
riplatesc

7H Solutions

Coming soon

What's it all about?

We all need water to stay alive. Some people like bottled mineral water. Others are happy with tap water. But is either of these liquids pure water?

Look at this label from a bottle of mineral water.

In this unit you will find out more about dissolving solids to make mixtures. You will also find out about separating these mixtures.

What do you remember?

You already know about:

- some solids that dissolve and others that do not
- separating mixtures of solids and liquids
- some liquids that do not contain water
- materials being made up of very small particles

1 Which one of these solids does NOT dissolve in water?

sugar salt chalk copper sulphate

2 How would you separate the solid bits from a mixture of soil and water?

evaporation filtration condensation freezing

3 Which one of the following does not contain water?

petrol milk cola vinegar

Ideas about solutions

QUESTIONS

Look at the cartoons above:
Discuss these questions with your partner.

a) What do you think has happened to the salt in the pan?

b) Do you think you can purify dirty water? How would you try to do it?

c) Why can't Pete get the solid back by filtering?

d) Will Molly and Pete really make the sweetest drink ever? Why?

e) Is it true that a substance which won't dissolve in water won't dissolve in any liquid? How would you help Mike get rid of the stain?

f) What is the difference between melting and dissolving?

Learn about

- finding out which liquids are pure and which are mixtures
- separating salt from rock salt

Water everywhere

Some people think that rain is the purest water you can get. Others might say that it's water from a mountain stream. But lots of substances, including some gases, dissolve in water.

Pure or mixture?

- You will be given two unknown liquids, labelled A and B. One of them is a pure substance and one is a mixture of a solid in a liquid.

How can you find out which liquid has the solid dissolved in it?

Show your plan to your teacher before starting any tests.

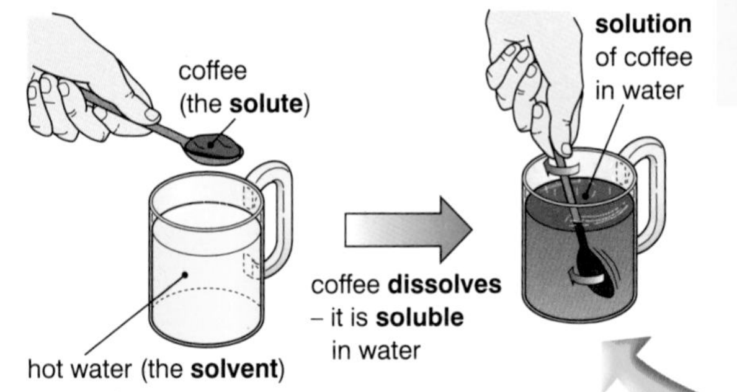

Definitions

When we look at substances dissolving we need to know the meanings of the words below.

Soluble: describes a substance that dissolves in a particular liquid.
Insoluble: describes a substance that does not dissolve in a particular liquid.
Solvent: the liquid that does the dissolving.
Solute: the substance that has dissolved in the liquid (solvent).
Solution: the mixture of solvent and solute.

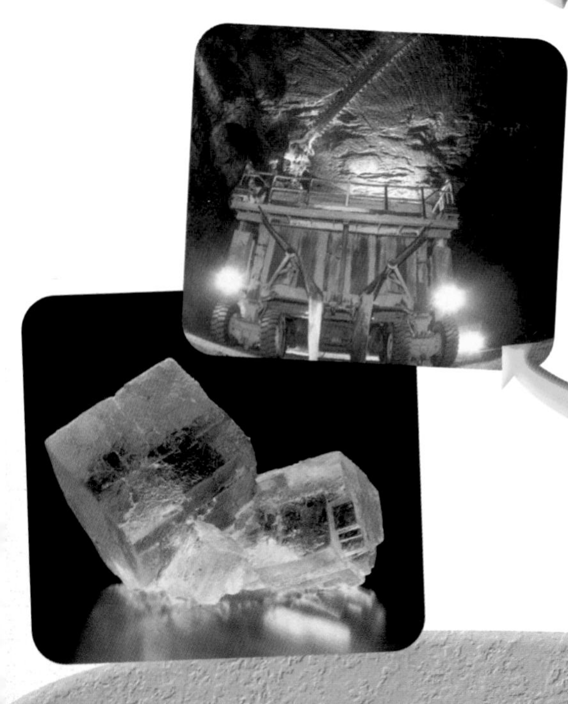

Rock salt

The salt that we sprinkle on food is called **sodium chloride**. We find it in nature as seams of rock salt under the ground. It is also the main solid dissolved in sea water. As you know, sodium chloride is soluble in water.

The sodium chloride in rock salt is mixed with sandy bits of rock. The bits of sand and rock are insoluble in water.

Getting the salt from rock salt

You can separate pure salt (sodium chloride) from rock salt. Follow the steps below to get a pure sample of salt.

ICT CHALLENGE

Find out how we can get salt from under the ground without digging a mine.

STEP 1

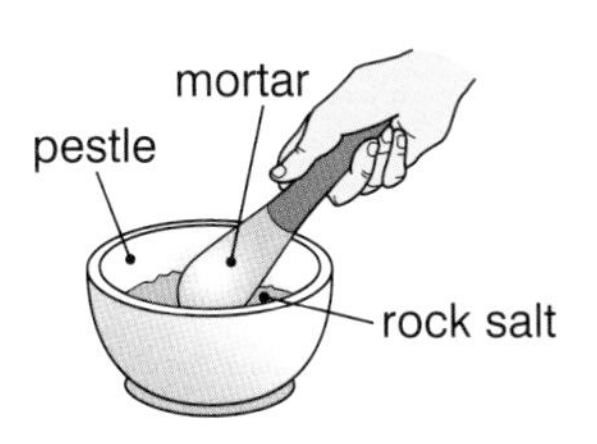

• Crush some rock salt

STEP 2

• Add to water and stir

STEP 3

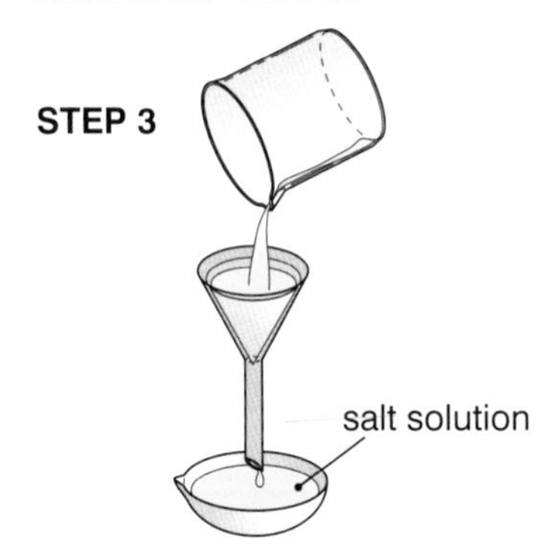

• Filter and collect the salty water (the filtrate)

STEP 4

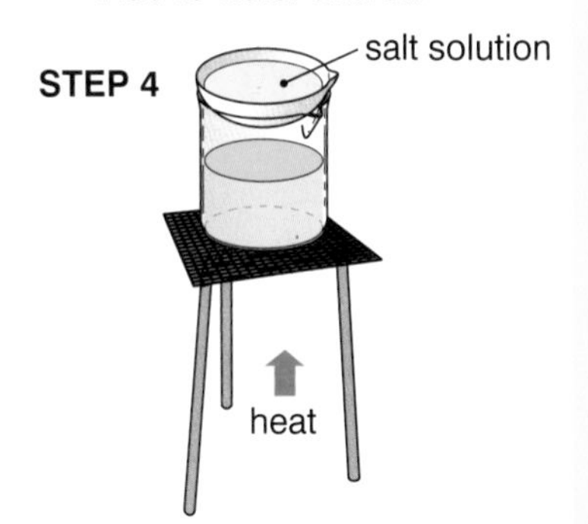

• Evaporate off half the water from the salt solution

STEP 5

• Leave the remaining solution until next lesson. Use a hand lens to look at the salt **crystals** left.

SUMMARY QUESTIONS

1 ☆ Copy and complete the sentences, using words from the list below.

solute dissolve solution solvent

A soluble solid will . . . in a liquid to form a . . . .
The soluble solid is called the . . . and the liquid is called the . . . .

2 ☆☆ Draw a flow diagram to show how you can separate salt from rock salt.

Key words

crystals
insoluble
sodium chloride
soluble
solute
solvent

7H2 Particles in solution

LEARN ABOUT
- conservation of mass
- using particle theory to explain dissolving

You know about the particles in solids and liquids from Unit 7G. But how can we use this model to explain dissolving?

Role play – Time to dissolve!

- Imagine that most of the students in your class are particles of water. They could wear something blue.

The rest of the students are the particles in a solid, such as salt. They could wear something white.

Now act out what you think happens when salt dissolves in water.

- Write down what you did.

Weigh the water and salt before mixing them together

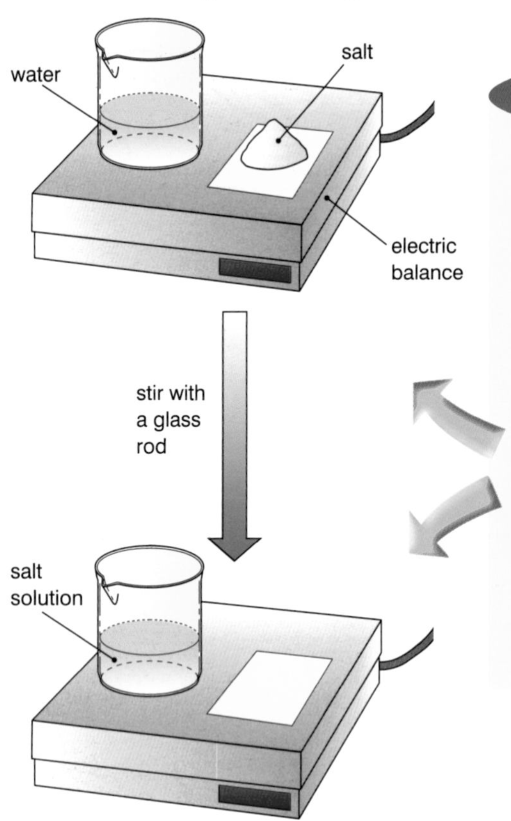

Weigh the water and salt after mixing them together

What happens to the mass?

- Think about your model of dissolving from your role play above.
  - **a)** What do you think will happen to the mass of salt and water before and after we stir them together? Will the total mass increase, decrease or stay the same?
- Follow the instructions in the pictures.
- Record the mass of the water, salt and apparatus before and after mixing.
  - **b)** Did your results support your prediction? Explain your evidence.

Q1 Why can't we get the salt back from salt solution by filtering it?

Q2 How can we get the salt from salt solution?

How solids dissolve

As a solid is added to a solvent, its particles are still in their usual regular arrangement.

The solvent particles are attracted to the particles in the solid.

The solvent particles 'pull at' the particles of the solute.

The particles of the solute get pulled away from their neighbouring particles in the solid.

As you know, the particles in a liquid are constantly moving around. They are slipping and sliding over each other. This makes the dissolved particles spread out evenly in the solution formed. The particles of the solute and the solvent **intermingle**.

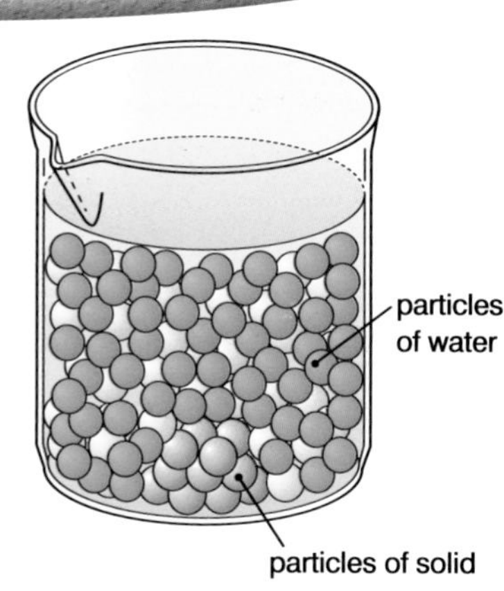

The solid before it dissolves

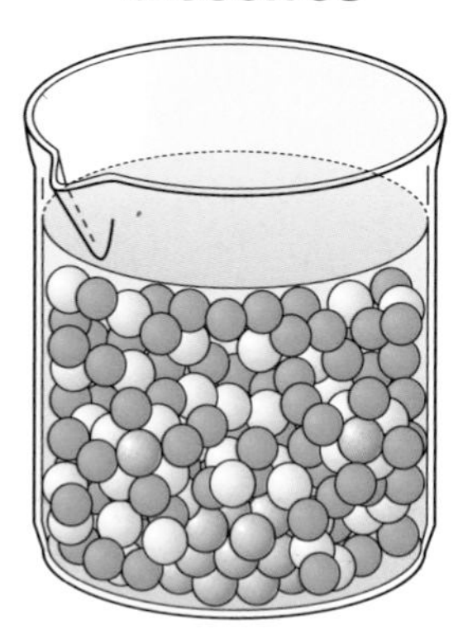

The process of dissolving begins

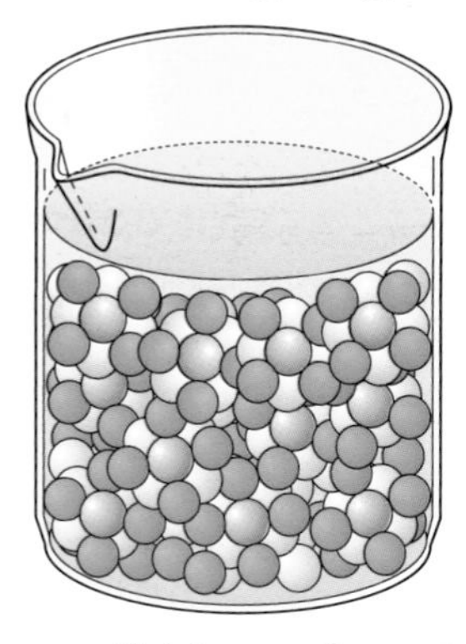

The solid has dissolved to form a solution

Investigating the rate of dissolving

- The question is:
  Which factors affect how quickly a solid dissolves?
- You can use copper sulphate as the solute and water as the solvent.

Your teacher will give you a sheet to help you investigate the question.

Let your teacher check your plan before you start.

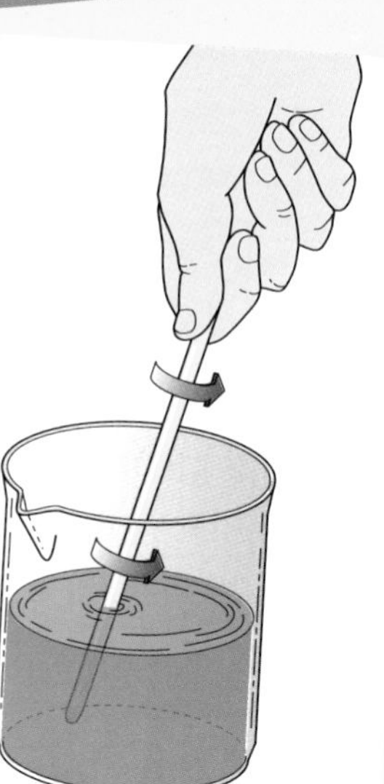

SUMMARY QUESTION

1 ☆ Copy and complete the sentences, using words from the list below.

solute particles attracted spread solvent

The particles of a solvent are quite strongly . . . to the . . . in a solute. The particles of . . . get 'pulled' from the other particles in the solid.

The movement of the particles of . . . results in the solute getting . . . throughout the solution.

Key words

attracted
intermingle

7H3 Distilling mixtures

LEARN ABOUT

- separating a solution by distillation
- explaining how distillation works.

A lavender field

The process of condensation isn't always useful; but it is when we distil a mixture

Uses of distillation

Have you ever brushed against a lavender bush? It smells wonderful. Its scented oil is used in aromatherapy treatments to help people relax.

We use a process called **distillation** to extract the perfume from the lavender bush.

When we distil a solution we can separate off the liquid and collect it. The distillation follows the steps below.

Distillation

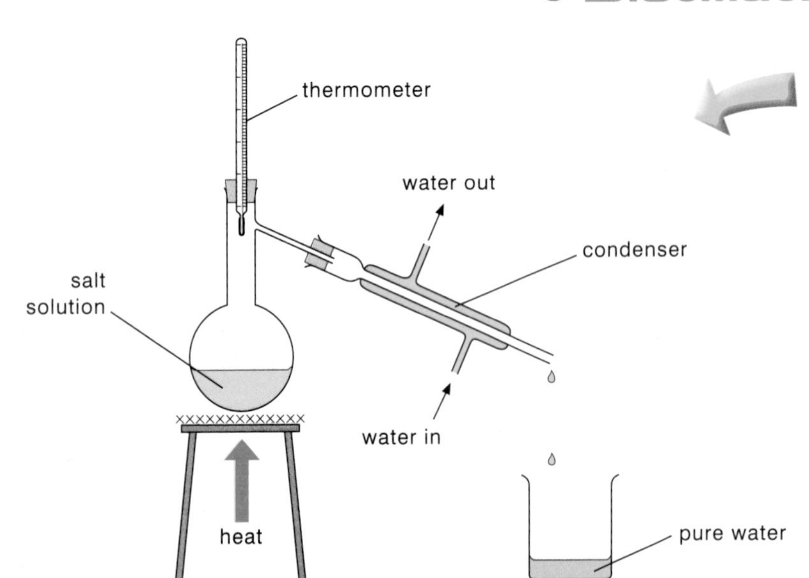

Look at the apparatus we use to do a simple distillation using a condenser:

- First of all, we heat the solution until the liquid boils.
- The gas given off is then cooled down. This usually happens in a condenser.
- This makes the gas condense and turn back into a liquid.
- The pure liquid can now be collected.

Any solids dissolved in the liquid remain in the heated flask.

Separating water from a solution

Try this experiment to see if you can separate water from a solution.

- Add a few drops of food colouring to 20 cm^3 of water in a boiling tube.
- Drop in a few anti-bumping granules. These will help the solution boil more smoothly.
- Set up the apparatus shown in the picture.

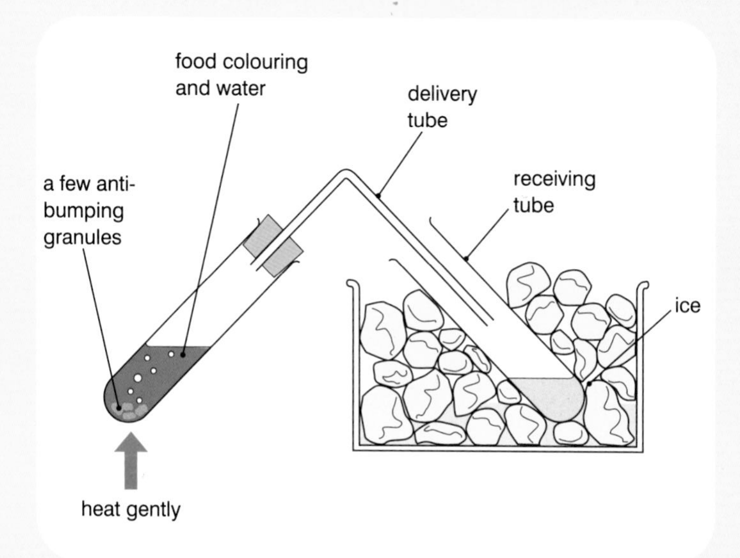

- Then heat the mixture **very gently**. You don't want the coloured mixture to shoot through the delivery tube.

  a) What collects in the receiving tube?

  b) Explain how your experiment worked. Use the words '**evaporated**' and '**condensed**' in your answer.

Scientists have identified over 280 different substances in the smell of a strawberry. That's why food scientists have struggled to make a really good artificial strawberry flavouring.

SUMMARY QUESTION

1 ☆ Copy and complete the sentences, using words from the list below.

liquid distillation condenser boil liquid cool

We can use the process called . . . to collect a pure . . . from a solution. When we distil a solution, we . . . it, then . . . down the gas given off in a . . . . The gas turns into a . . . which we collect.

Key words

condense
distil
distillation
evaporate

Chromatography revealed

Learn about

- separating solutes by chromatography
- interpreting chromatograms

Sweet chromatography

Do you know what you are eating when you suck on a coloured sweet? How do the manufacturers make those bright colours? Do they have a single dye for each colour? Or is each colour made by mixing different dyes?

Chromatography can help us to find the answers.

You can use chromatography to separate two or more solutes that are dissolved in a solvent.

Dyes in inks

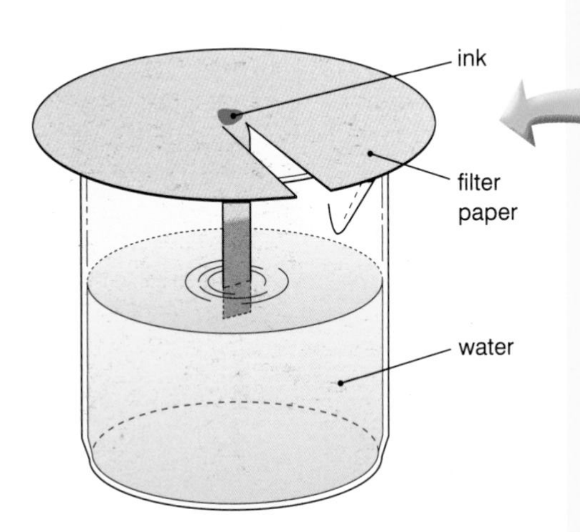

- Set up the experiment shown in the picture. Let the water soak up the wick. It will spread out on to the filter paper.
- You can try different-coloured water-soluble inks.
  a) Why do the inks have to be water soluble.
  b) Which inks are made from only one dye?
  c) Which inks are mixtures?
  d) Which colours are mixed to make each ink?
  The pieces of paper left at the end of the experiment are called **chromatograms**.
- Let your favourite chromatogram dry and stick it in your book.

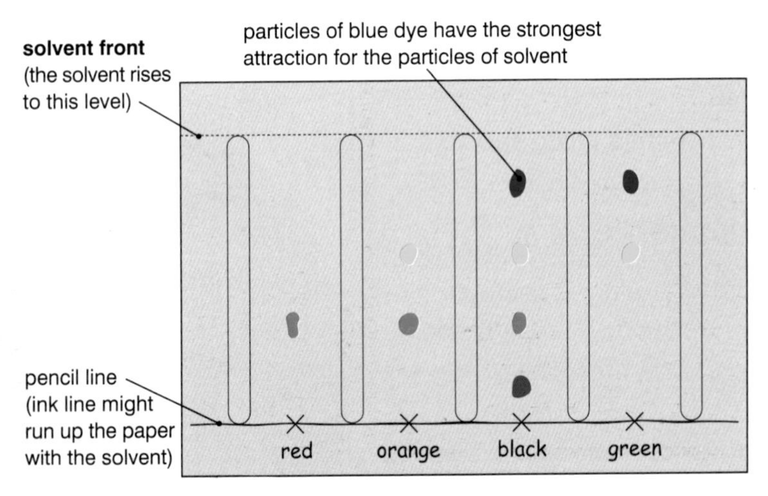

You can also make chromatograms by letting the solvent run up absorbent paper. In this case you dip the paper into the solvent. The different solutes are carried up the paper for different distances. Look at the example in the picture.

Testing sweets

In this experiment you can test the dyes used to colour sweets.

To make a good chromatogram you need to start with small, concentrated dots of each colour.

Your teacher will give you a set of steps to follow.

What conclusions can you draw from your chromatogram?

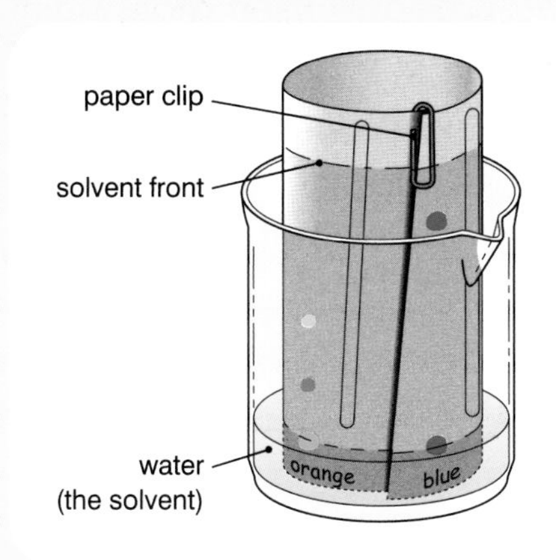

Using chromatography

Forensic scientists use chromatography to identify unknown substances from the scene of a crime.

It can also show if someone has tried to forge your signature. The forger can write in ink that seems to be the same colour as the ink that you used. However, chromatography will show up any differences between your ink and the ink used by the forger.

SUMMARY QUESTION

1 ☆☆ Copy and complete the sentences, using words from the list below.

separate attraction solvent attraction furthest solvent chromatography

. . . is used to . . . a mixture of solutes. You use a . . . in which the solutes will dissolve. There will be different forces of . . . between the particles of the . . . and the particles of solute. The solutes with the strongest . . . will be carried along . . . by the solvent.

Key words

chromatogram
chromatography
solvent front

Learn about

- solubility in different solvents
- solubility curves

Which colour would you choose for your bedroom?

Solvents

Have you ever helped to decorate a room? If you have painted a door, you probably used gloss paint. At the end of the job you have to clean the paint off your brush.

Gloss paint is an oil-based paint so water can't dissolve it. As you probably know, oil and water do not mix. You have to wash your brush in white spirit – a liquid that does dissolve oil.

We say that water and white spirit are both good solvents, but for different substances.

Looking at different solvents

You will be given two solvents to investigate – water and ethanol.

Your task is to answer the question:

How much solid will dissolve in each solvent?

When no more solid will dissolve, we have made a **saturated** solution.

The solids you can use are sodium chloride and potassium nitrate.

Your teacher will give you a sheet to help.

SAFETY: Ethanol is highly flammable so make sure that no Bunsen burners are alight when you use it.

- Record your results in a table like the one below.

Solvent	Mass of sodium chloride that dissolves in 20 cm^3 of solvent	Mass of potassium nitrate that dissolves in 20 cm^3 of solvent
Water		
Ethanol		

What is your conclusion?

No room for us... this pool is saturated.

Saturated solutions

Only a certain mass of solute can dissolve in a particular solvent at a given temperature. This mass varies from solute to solute.

Q1 How could you test whether a solution is saturated?

We can describe how well a solute dissolves in water by its **solubility**. The solubility of a substance tells us how many grams of it will dissolve in 100 g of water. We must also say at what temperature. For example, the solubility of sodium chloride is 36 g per 100 g of water at 25°C.

Q2 Look at the solubility of sodium chloride above. Why do you think we must state the temperature?

We can find the solubility of a solid at different temperatures. We observe when crystals first appear from a cooling solution. At this point the solution is saturated.

When we have several results, we can plot a graph. It shows how the solubility changes with temperature. This is called a **solubility curve**.

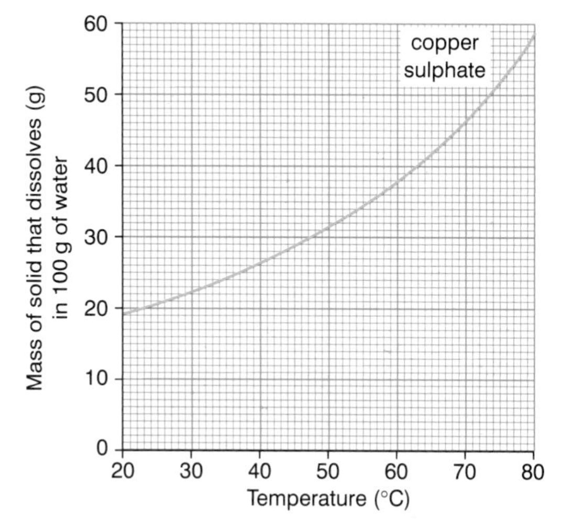

The solubility curve of copper sulphate

SUMMARY QUESTIONS

1 ☆ Copy and complete the sentences, using words from the list below.

water solubility curve temperature solubility

The . . . of different solutes will vary in different solvents.
. . . also affects the . . . of solutes.
We quote the solubility of a solute in g per 100 g of . . . at a certain temperature.
We can show how solubility changes with temperature on a solubility . . . .

2 ☆☆☆ The solubility of potassium sulphate in water increases as the temperature rises.

a) Put these data into a table:
At 10°C, 9.3 g dissolve; at 30°C, 13 g dissolve; at 50°C, 16.5 g dissolve; at 70°C, 19.8 g dissolve; at 90°C, 22.9 g dissolve.

b) Draw a graph to display the data. (Solubility goes up the side and temperature along the bottom.)

c) Use your graph to find the solubility of potassium sulphate at 20°C.

Key words

saturated solution
solubility

7H
Read all about it!
IDEAS AND EVIDENCE

Chromatography at work – Dopey horses

Horse racing is a multi-million pound industry. With so much money at stake, some people are willing to break the rules of the sport. One way to 'fix' the results of a race is to give a horse some kind of drug.

Some drugs act as stimulants and help a horse run faster. Other drugs 'dope' a horse and have the opposite effect, slowing it down. In this way, criminals gamble on a race and try to make sure their bet wins.

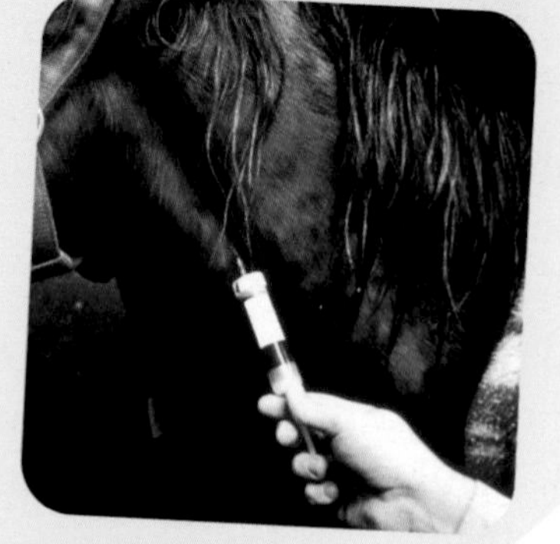

Trainers of an injured horse can be tempted to give it painkilling injections so it can run in a big race. This could permanently damage the horse.

That's why we need scientists to check that horses are free from drugs. They use a technique called **gas chromatography**. This separates the drug from natural substances in the horse's urine or blood.

 SCIENTIFIC PEOPLE 

Drug testers

Horse racing isn't the only sport that checks for drugs. Drugs have hit the headlines in football, tennis, athletics, cycling and swimming.

Some drugs help the cheats to build muscle. Others help you to train for longer and harder. Those determined to gain an unfair advantage risk their own health taking these drugs.

The drug takers go to any lengths to avoid getting caught. A new drug that masks an illegal drug has been developed. However, drug testers have found a way to detect it. So athletes who thought they were safe have recently been caught and banned from the sport.

The drug testers can visit a sportsperson at any time – not just when they are in competition. Their instruments are also getting more sensitive. So the battle against the cheats goes on.

England footballer, Rio Ferdinand, got into trouble for missing a drugs test in 2003. A later test found he was clear, so why do you think he still received a ban?

- When a solid dissolves in a liquid their particles intermingle.
- The solid is called the **solute**. The liquid is the **solvent**. The mixture we get is called a **solution**.
- We can collect the solvent (liquid) from the solution by **distillation**.
- If a solvent contains two or more solutes, we can separate the solutes by **chromatography**.
- A solution that will not dissolve any more solid at a particular temperature is called a **saturated** solution.
- The **solubility** of a substance varies with temperature. We can show this on a graph called a **solubility curve**.

DANGER! AVOID THESE COMMON ERRORS

When a solute dissolves it only *seems* to disappear – its particles are still there. The particles intermingle with the particles of the solvent.

A solute will dissolve only until the solution becomes saturated at a particular temperature. At this point you will see solid at the bottom of the solution if you add any more solute.

Key words

chromatography
distillation
saturated
solubility
solute
solution
solvent

UNIT REVIEW

REVIEW QUESTIONS

Applying and understanding concepts

1 Copy and complete the sentences, using words from the list below.

solubility liquid chromatography grams solute distillation saturated solvent

The solid that dissolves in a . . . to form a solution is called the . . . .

The liquid that does the dissolving is called the . . . .

A solution that cannot dissolve any more of the solute is said to be a . . . solution.

The . . . of a solute at a certain temperature is measured in grams per 100 . . . of water.

We can separate and collect a liquid from a solution by . . . .

We can separate different solutes from a solution by . . . .

2 Food scientists test a new type of orange squash to see if it contains a banned substance. The scientists have three banned dyes that they think might be used to colour the squash and two other dyes that are safe to use. Look at their paper chromatogram.

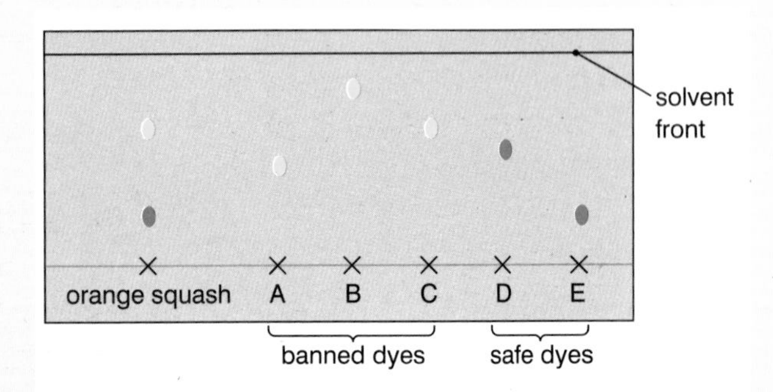

a How many dyes are in the new squash?

b Which banned dye does the squash contain? Explain how you worked this out.

3 Draw a concept map linking these terms together. Don't forget to label your links, explaining what the connection is.

water sodium chloride distillation rock salt filtration

Ways with words

4 Imagine you are a particle of sugar about to be added to a cup of tea. Write a story of your adventure as you travel to a hotter place!

Making more of maths

5 A group was finding out the proportion of salt in rock salt. They started with 50 g of rock salt and prepared 29 g of salt.

What was the percentage of salt in the rock salt?

6 Look at the solubility curves of copper sulphate and sodium chloride on the graph.

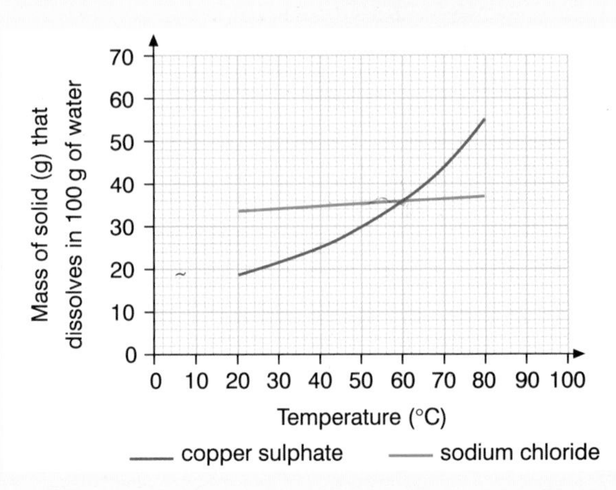

a What is the solubility of copper sulphate at 40 °C?
b What is the solubility of sodium chloride at 60 °C?
c In which of the two solids is solubility affected more by temperature?
d At which temperature do both solids have the same solubility?

Extension question

7 Using students in yellow and blue bibs from PE, explain how you could model:
a a solution of sugar forming as it dissolves
b the distillation of salt solution.
Draw diagrams if this will help.

SAT-STYLE QUESTIONS

1 The apparatus shown below is used to separate pure water from impure water.

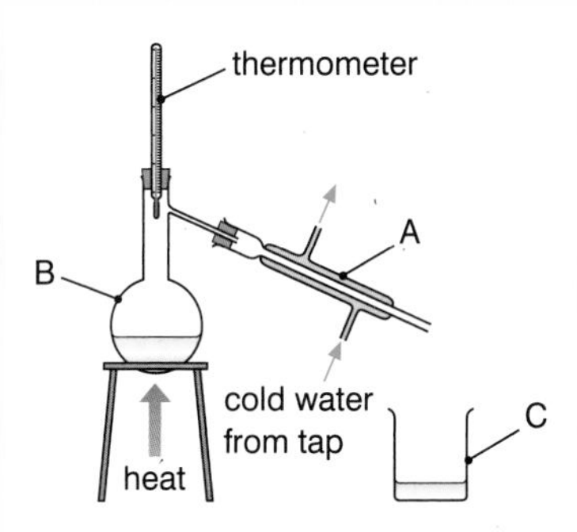

a What would be the temperature on the thermometer? (1)
b Where would you place the impure water in the apparatus? (1)
c Where would the pure water collect? (1)
d What is the function of A? (1)
e What do we call this process? (1)

2 Sam made a chromatogram using four different felt-tipped pens. His chromatogram is shown below.

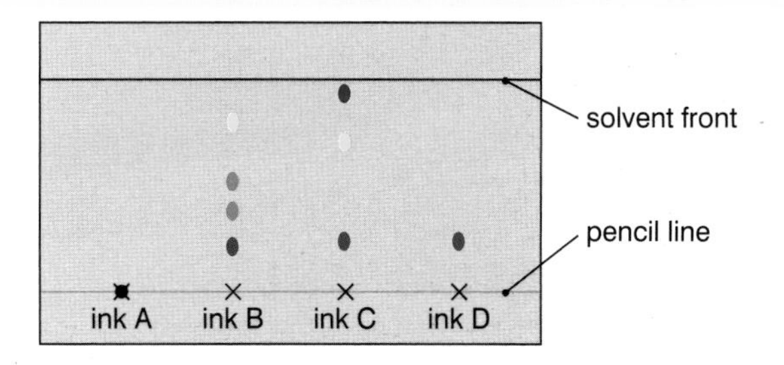

a Why did Sam use a pencil, and not ink, to draw a line across the bottom of the paper? (1)
b Which ink was a mixture of four dyes? (1)
c Which ink definitely contained only one coloured dye? (1)
d Which ink had the weakest attraction for the solvent? (1)
e Which ink contained the dye with the strongest attraction for the solvent? (1)

3 Anna wanted to see which of two solids, A or B, was more soluble in water and in ethanol.
She decided to count how many spatulas of each solid would dissolve.
Here are Anna's results:

Solid	Number of spatulas dissolved	
	in water	in ethanol
A	5	3
B	1	4

a Name one thing Anna had to do to make this a fair test. (1)
b Which solid was more soluble in water? (1)
c What is the missing word? Water and ethanol are . . . for A and B. (1)
d Which two results would you check again? Why? (1)

Key words

Unscramble these:
tosuel
velnots
tidlis

COMING SOON

What's it all about?

Have you ever climbed a mountain? Or swum a mile? It's tiring doing things like this and you soon run out of energy. You can renew your energy reserves by eating food.

Energy is something you need whenever you are doing something. Even sitting and thinking uses up energy.

Scientists often use the idea of energy when they are talking about different activities. In this unit, you can find out about the scientific idea of energy. You will think about how we use energy resources, and why we may soon have to change some of the things we do.

What do you remember?

You already know about:

- how plants grow
- materials that burn
- keeping things warm

1 Which of these things does a green plant need if it is to grow?
light water warmth air

2 Name some materials which can be burned. What substance from the air is needed for burning?

3 Which can be used as a fuel?
air water oil sunlight

4 What do we call a material that is good at keeping in heat? Give some examples.

Young technologists

QUESTIONS

Lots of scientists and technologists are working to design better cars – cars that use less fuel. Some are trying to design cars that don't use any fuel at all.

- What fuels do most cars use?
- A few cars run on electricity. Where do they get their electricity from?
- Look at the car in the picture above. It doesn't use fuel. So what different things are supposed to make it go? Do you think a car like this could ever work?
- Why would it be good to design a car that runs on sunlight? Can you think of any disadvantages of a 'solar car'?

Fuels on fire

Learn about
- burning fuels
- energy from fuels

Burning bright

Around the world, many millions of people use wood or charcoal as their **fuel** for cooking and heating. Fuel is important. It's often the job of children to gather wood, and they may spend several hours a day at this work.

A fire releases the **energy** stored in the fuel.

Q1 What other ways do we have of heating houses?

Food can taste really good if you cook it out of doors, on a barbecue. The fuel is charcoal or gas.

What is a fuel?

A fuel is a material which we burn in order to release the energy that it stores. There are many different fuels in use. There are even power stations which burn household rubbish to generate electricity.

Q2 Name some fuels which are used for cooking.

Testing Bunsen flames

Have you learned how to use a Bunsen burner safely? You can use a Bunsen burner to find out about the energy released when gas is burned.

- Try heating some water in a glass beaker. Use a thermometer to see the temperature rise.
- Compare your results with other people's results. How can you be sure that your comparison is as fair as possible?

SAFETY Always wear eye protection when using a Bunsen burner.

Useful fuels

When fuels burn, their energy is released. So what do we mean by 'energy'? The pictures will give you some ideas.

light

fuel = wax

You can't see the energy stored in a fuel, but you can certainly tell when the energy is released. You get light, heat, movement or electricity. We can think of all of these as 'energy on the move'.

This bike uses petrol as its fuel. When its energy is released, it can make the bike move very quickly.

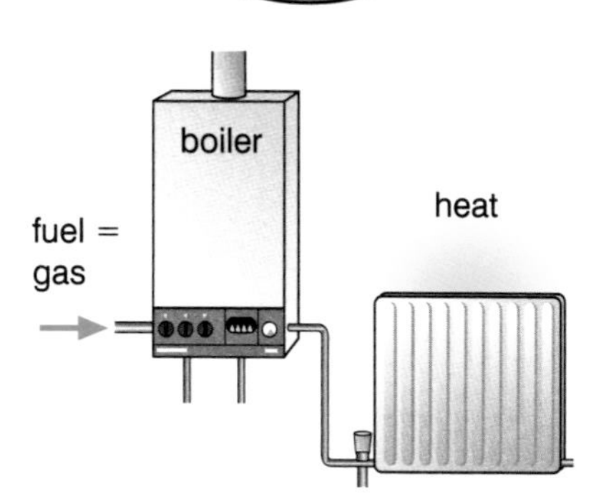

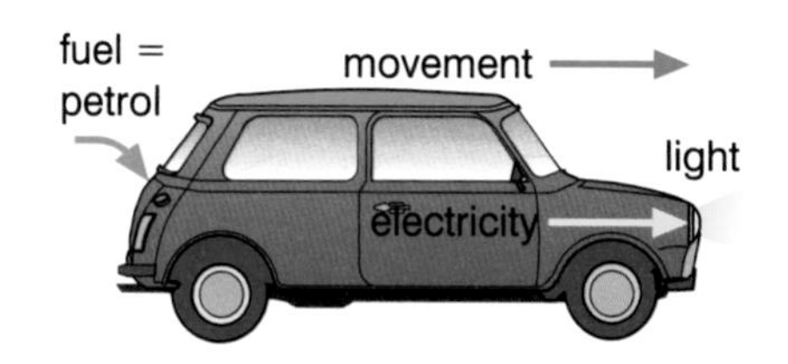

More fuels

Here are some more examples of fuels at work:

- a petrol-powered lawnmower
- a bonfire
- a gas lamp, used for camping.

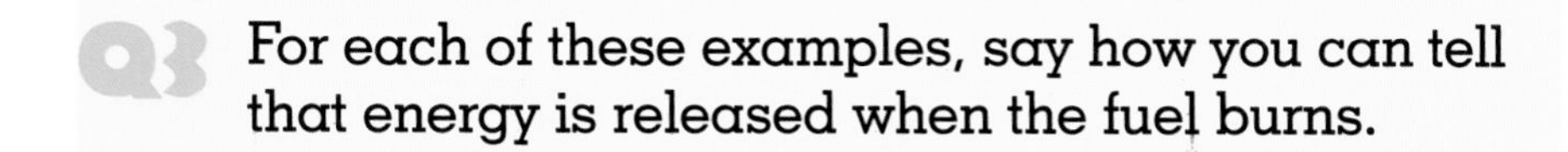

Q3 For each of these examples, say how you can tell that energy is released when the fuel burns.

ICT CHALLENGE

Does petrol really 'burn' in a car engine? Use the Internet or a CD-ROM encyclopaedia to find an animation of how petrol is used in a car engine.

SUMMARY QUESTIONS

1 ☆ Skim through these two pages. As you do so, list all the different fuels which are mentioned. Can you add any more fuels to the list?

2 ☆ What do we have to do to release the energy of a fuel?

3 ☆☆ Find out which fuels are used in these different situations:

camping stoves aircraft hot air balloons
heating the water in a swimming pool

Fossil fuels

Learn about
- using fossil fuels
- how they were formed

Much of the gas used in the UK for cooking and heating comes from under the seas around us. A giant network of pipes carries it around the country.

Going up in smoke

Coal, oil and natural gas – these are **fossil fuels**. They are extracted from underground. Coal is mined, but oil and gas are usually brought to the surface through pipes. Oil is usually transported around the world in pipelines and tankers.

Generating electricity

Sometimes we use the energy from fossil fuels without even thinking about it. Most of the electricity we use in the UK comes from fossil fuels. Coal and gas are burned in the **power stations** where electricity is generated.

Q1 Explain how oil is important for transport.

This big power station has a giant stock of coal. It needs a trainload to be delivered every hour if it is to keep generating at full power.

The story of fossil fuels

Once upon a time, 300 million years or so ago, Britain was rather different. The climate was warm, and swampy forests grew. There were ferns and mosses, reptiles and amphibians, but no mammals.

When plants died, they fell into the swamps. At first, they rotted away to become peat. Then they became buried deeper. Heat and pressure squashed them until they eventually became coal.

Oil and gas are formed in a similar way, but from plant and animal remains in the sea.

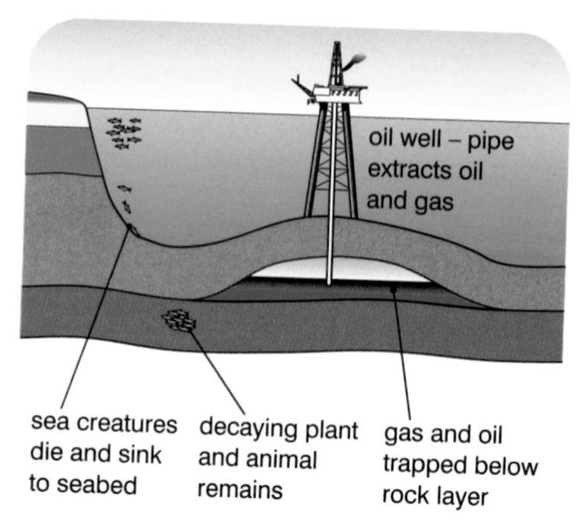

Oil and gas try to rise up from underground, but they get trapped by solid rock

Running low

Once fossil fuels have been burned, they are gone forever. Remember that they formed hundreds of millions of years ago – and no-one wants to wait that long when their car runs out of petrol!

- List some everyday activities which make use of fossil fuels.
- For each one, suggest how we might be able to continue with the activity without using fossil fuels.

Big problems

Fossil fuels are very useful to us. But there are problems. When fuels are burned, they release gases. These cause **pollution** and **climate change**.

AMAZING SCIENCE!

Peat is a fossil fuel. In Ireland, there is a power station which burns nothing but peat.

Q2 How is climate change likely to affect the temperature of the Earth? What other problems have you heard of that might be caused by climate change?

SUMMARY QUESTIONS

1 ☆ Name as many fossil fuels as you can.

2 ☆ Explain why wood is a fuel, but not a fossil fuel.

3 ☆☆☆ Draw a chart to show how the energy we get from burning coal came from plants that lived millions of years ago.

Key words

climate change
fossil fuel
pollution
power station

7I3 Renewable energy

LEARN ABOUT

- renewable energy resources

Energy from trees

We will eventually run out of fossil fuels. We need to find other **energy resources**. An energy resource is anything from which we can get energy.

In the UK, farmers are experimenting with fast-growing trees. After a few years, these are cut down and used as fuel in power stations.

This farmer in Oxfordshire is growing willow trees as a renewable energy resource

Energy for ever

We need to find energy resources that can be replaced as quickly as we use them up – they must be **renewable** energy resources. Trees are renewable because we can grow new trees to replace the ones we cut down.

We describe fossil fuels as **non-renewable** energy resources. Once coal, oil and gas are burned, we can't get them back again – unless we wait for millions of years.

This taxi runs on bio-diesel made from maize

Q1 Is bio-diesel a renewable or a non-renewable energy resource? Explain your answer.

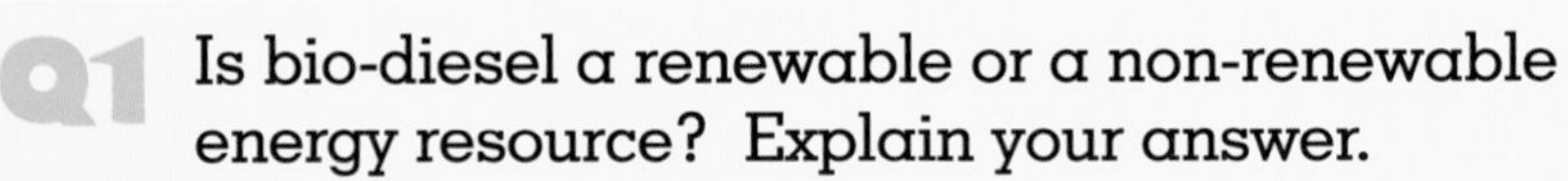

Electricity from renewables

- Try out some model systems which show how electricity can be generated from:
  - **a)** sunlight
  - **b)** moving water
  - **c)** moving air.
- For each model, suggest how a full-sized version is used for generating electricity.

A supermarket in south Wales sold out of vegetable oil because people were using it as a cheap fuel in their cars!

Cleaner electricity

Electricity is a clean and easy way of transferring energy from place to place, and we use a lot of it. Here are some ways of generating electricity from renewable resources:

- **hydroelectric power** – using moving water
- **wind power** – using moving air
- **wave power** – using the movement of waves on the sea
- **solar power** – from sunlight, using solar cells
- **geothermal power** – from hot rocks inside the Earth.

You can sense the energy of the water gushing from this hydro-electric dam in Canada

Q2 Explain why each of these energy resources can be described as renewable.

No harm done?

Renewable energy resources sound good, but they're not perfect. Each comes with its own problems. They don't produce gases which damage the atmosphere, but they are usually expensive.

For example, a hydroelectric dam floods land, and people may have to move away. Wind farms may be noisy, and some people think they are ugly.

Q2 Can we rely on electricity from renewable energy sources? Discuss what would happen on a day without wind or sunshine.

I guess they're called wind generators because they're used to generate wind.

SUMMARY QUESTIONS

1 ☆ Skim through these two pages. As you do so, list all the different renewable energy resources which are mentioned.

2 ☆☆ Is natural gas a renewable or a non-renewable energy resource? Explain your answer.

3 ☆☆ Most spacecraft use solar cells to generate electricity. Explain why this is a good choice.

Key words

energy resource
non-renewable
renewable

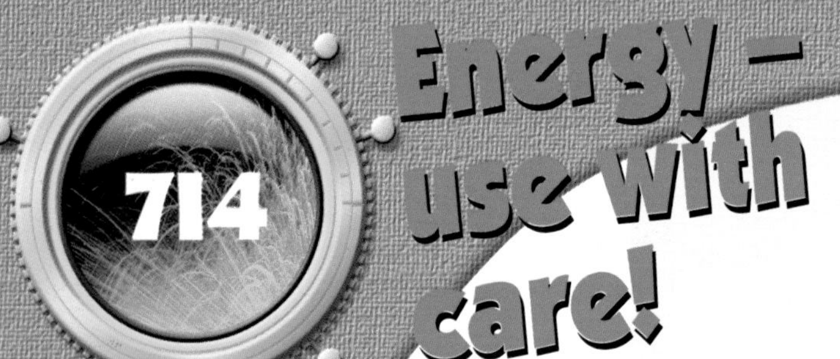

7I4 Energy – use with care!

Learn about
- conserving energy resources

Buildings old and new

Many blocks of flats like this have had double-glazed windows fitted to help save energy

Many old buildings were built when people weren't very concerned about energy. Fuel was cheap, and they didn't realise that they were damaging the environment by burning fossil fuels.

The flats in this photo have very thin walls and lots of big windows. When the flats were built, the windows were single-glazed. That made it very easy for heat to escape.

Q1 Today, the law says that all new buildings must have well-insulated walls, floors and roofs. Their windows must be double-glazed. Why is this better for the environment?

Better by design

You could live comfortably in a place like this, and it's good for the environment

Look at the building in this photo. It has been carefully designed to use less energy. Its walls are good at keeping heat in.

There are other ways that builders can save energy. They can put in:

- big windows facing south, to catch the energy of the Sun's rays – it's free!
- smaller windows facing north (the cold side of the building), so that less heat escapes through them.

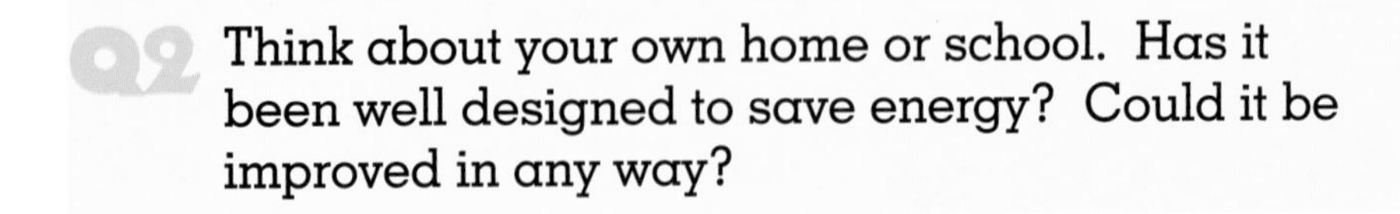

Q2 Think about your own home or school. Has it been well designed to save energy? Could it be improved in any way?

Conserving energy

You can see that architects and builders have to think carefully about energy resources. Energy is expensive. We are running out of fossil fuels, and we are harming the environment. We need to **conserve** our **energy resources** – by using them carefully and wasting as little as possible.

If you have solar cells on your roof, you can make money by selling any spare electricity to the supply company!

Capturing sunshine

Sunlight is a wonderful energy resource. There's a lot of it, and it's free.

- **Solar cells** change sunlight into electricity.
- **Solar panels** use sunlight to heat water.

Solar panels are like radiators in reverse. When sunlight falls on a panel, the water inside it gets hot. The hot water is useful for washing and for heating.

Crete is a sunny place, so it's a good idea to use the sunshine to heat water. You can see the pipes which take the hot water indoors.

Q3 Hot water stores energy for heating at night. Why is it useful to be able to store energy from sunlight?

Designing solar panels

What makes a good solar panel? The water inside it should heat up quickly when sunlight falls on it. Try an experiment to investigate some of the factors involved.

- Use an aluminium food tray. Put water in it and place it in the sunshine.
- Use a thermometer to find out how quickly the temperature of the water rises.
- What can you change? Think about the shape and size of the tray, the amount of water, and its colour. How about stretching plastic film over the top of the tray?

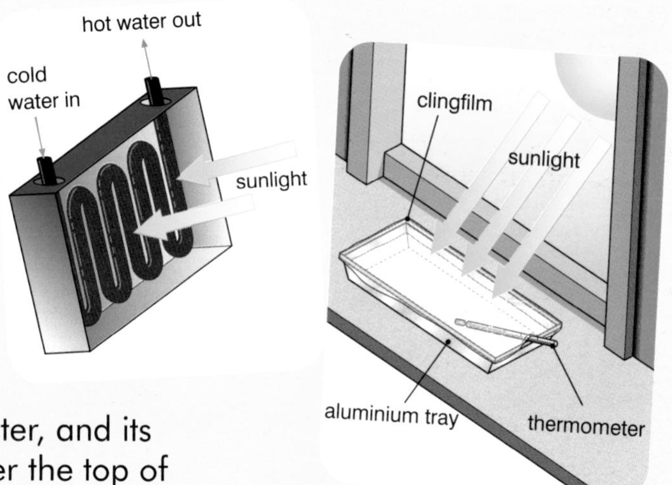

SUMMARY QUESTIONS

1 ☆☆ Give as many reasons as you can why we should conserve our energy resources.

2 ☆☆ How can houses be better designed to use sunlight as an energy resource?

3 ☆☆☆ Copy the diagrams and complete them to show how we can use the energy of sunlight.

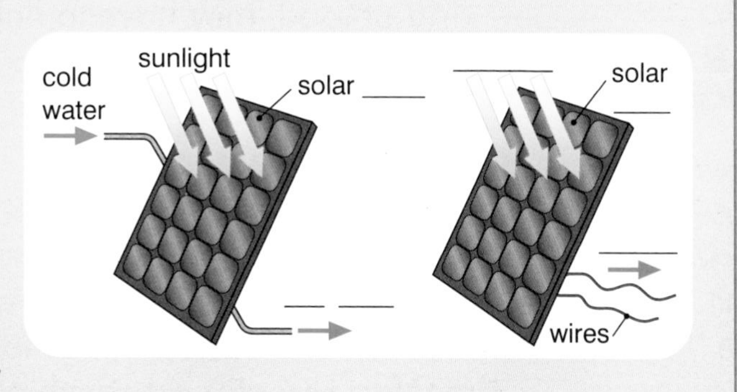

Key words
conserve
energy resources
solar cells
solar panels

Food as fuel

LEARN ABOUT
- energy from food
- the Sun as an energy source

Energy for activity

Human beings need energy all the time, especially if we are being active. Play a tough game of hockey or tennis and your body tells you that you need more energy, quickly – you feel hungry. **Food** is our energy resource.

Most packaged foods have a label to tell you how much energy they supply. Energy is stored in the carbohydrates, fats and proteins which make up food.

People concerned with their diet tend to think of food energy in calories. Scientists think in kilojoules.

Joules

The scientific unit of energy is the **joule**. This is sometimes written as **J**.

If you eat an apple, it will give you about 400 000 J of energy.

In a day, we need about 10 million joules of energy from our food.

Q1 Look at the food label in the picture. How much energy do you get from eating 100 g of baked beans?

Diet cola contains virtually no energy. You use more energy opening the can and lifting it to your mouth than you get from the drink.

Burning food

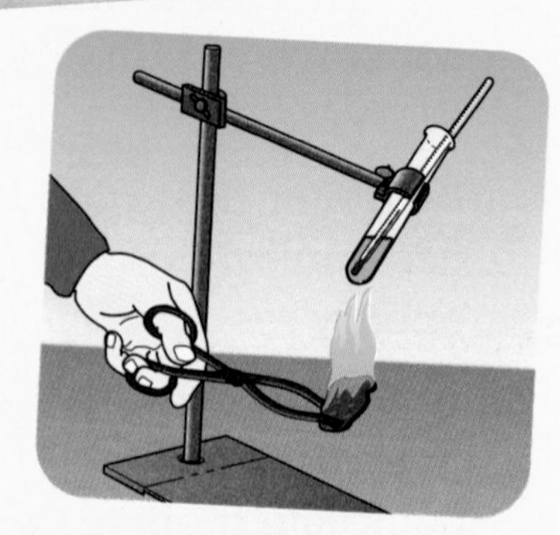

When food technologists invent a new kind of food, they have to find out how much energy it contains. They do this by burning it, so that it heats some water. The foods with the most energy make the water hottest.

- Think of an experiment to compare the energy resource in foods. The easiest things to burn are dry foods, like breakfast cereals and crispbreads.

Plant food

Where do plants get their energy? They get it from sunlight.

On a sunny day, plants make sugar and starch in their leaves.

The sugar and starch contain some of the energy of the sunlight.

These green leaves are trying to capture as much sunlight as possible

Energy from the Sun

Plants get their energy from the Sun. In fact, most of our energy originates from sunlight:

- **coal** – energy from plants that grew long ago
- **wind power** – energy from wind, caused by the Sun heating the air
- **water power** – energy from moving water. The heat of the Sun evaporates water, which later falls as rain.

Q2 Explain how the energy of oil and natural gas came originally from the Sun.

Gruesome science

Scientists have found ways of injecting water into mass-produced chickens so that each kilogram contains fewer joules of energy than it used to.

SUMMARY QUESTIONS

1 ☆ What is the scientific unit of energy?

2 ☆☆ Why do green plants have chlorophyll in their leaves?

3 ☆☆☆ Draw a cartoon strip to show how the energy we get by burning coal came originally from the Sun.

Key words

food
joule (J)
water power
wind power

71
Read all about it!
IDEAS AND EVIDENCE

Going under?

People on the islands of Tuvalu in the Pacific Ocean are very concerned. The sea is rising, and their islands are in danger of being flooded. They blame climate change, caused by everyone else using fossil fuels.

As the Earth's **temperature** rises, the polar ice caps melt and sea levels rise. The 11 000 Tuvalu islanders, who don't use much fossil fuel, feel that they are paying the price for everyone else's thoughtless **energy consumption**.

A tropical paradise under threat from climate change

Fair shares for all?

Some people use more energy than others. The diagram shows this.

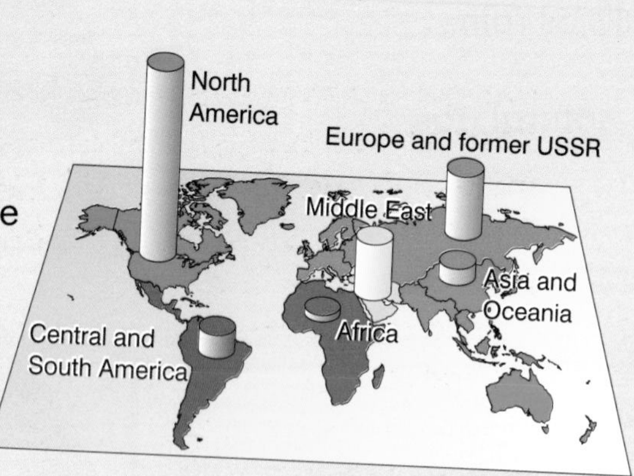

Q1 Study the chart. In which parts of the world do people use most energy? Which people use least energy?

SCIENTIFIC PEOPLE

Predicting climate change

The Earth's temperature is going up – it's official. The United Nations set up an international panel of scientists to decide whether our climate really was changing.

As new scientific **evidence** was gathered during the 1990s, it became clear that:

- temperatures are rising
- the culprit is probably **carbon dioxide**.

Q2 Why is there more carbon dioxide in the atmosphere these days?

Q3 Look at the graph. How do scientists expect temperatures to change in the future?

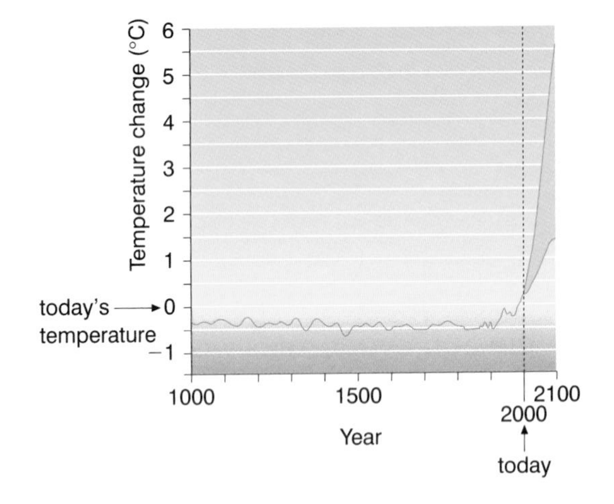

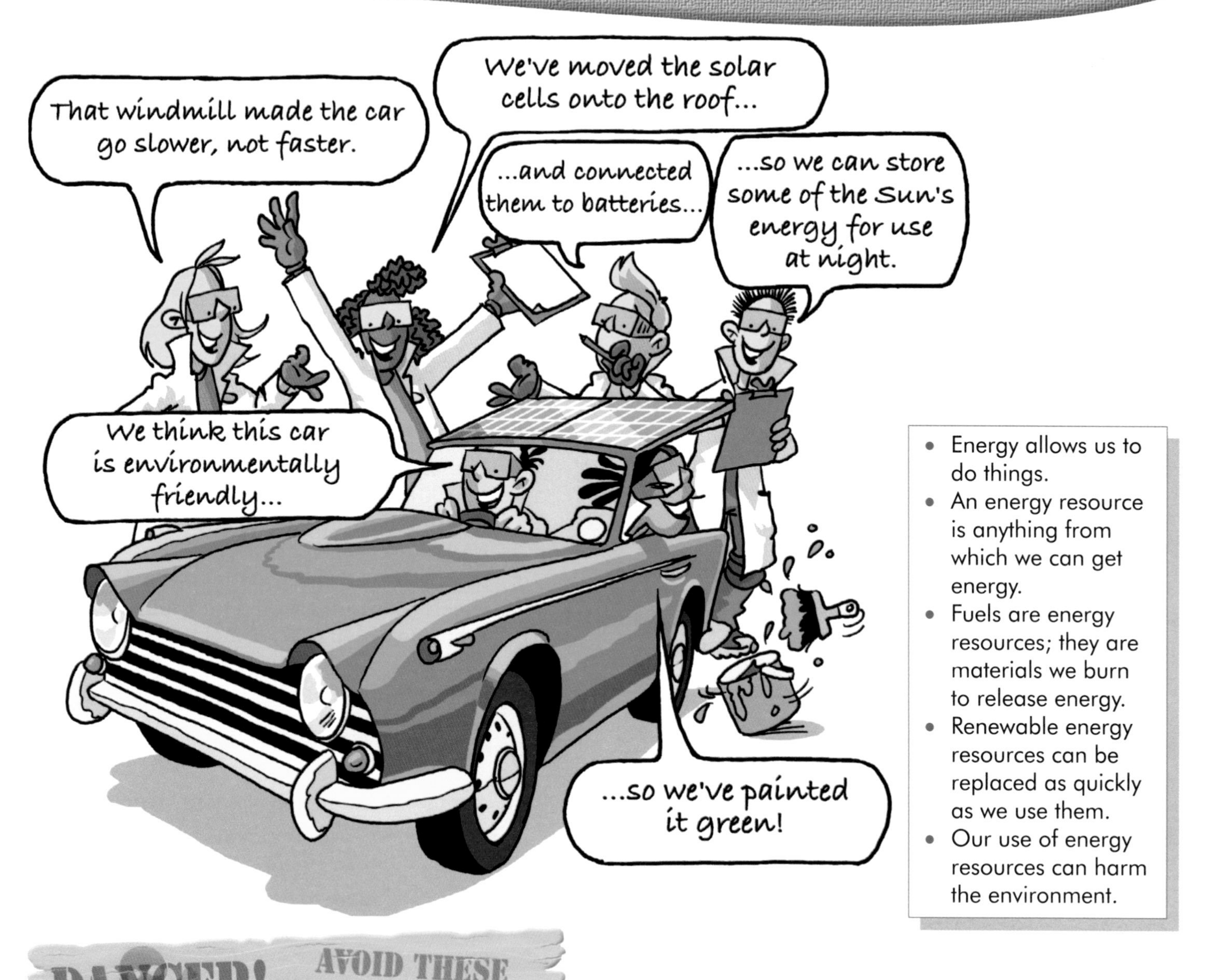

- Energy allows us to do things.
- An energy resource is anything from which we can get energy.
- Fuels are energy resources; they are materials we burn to release energy.
- Renewable energy resources can be replaced as quickly as we use them.
- Our use of energy resources can harm the environment.

DANGER! AVOID THESE COMMON ERRORS

How do you think of energy? It's something we need whenever we want to be active. We can get energy in lots of different ways – from food and fuels, from wind and moving water, and so on. These are our **energy resources**.

But be careful! Coal isn't energy. Wood and water aren't energy. Energy isn't stuff.

Think about burning some wood. There is a chemical reaction as the wood combines with oxygen from the air. We get heat and light. The wood has changed into ash, carbon dioxide and water vapour.

You should now have some idea of how to recognise energy resources. Water flowing downhill has energy. You can use that energy by making the water turn an electricity generator. Petrol is an energy resource, because you can burn it in a car's engine and make the car move.

Look out for energy resources all around you. The energy we get from them is what allows us to be active!

Key words

carbon dioxide
energy
consumption
evidence
temperature

UNIT REVIEW

REVIEW QUESTIONS

Understanding and applying concepts

1 We need energy to be active. What energy resources are used by each of the following? (There may be more than one answer in each case.)
  - **a** a rabbit
  - **b** a car
  - **c** a train
  - **d** a solar cell

2 We need to conserve energy resources.

Draw a picture to show a situation where lots of energy resources are being wasted. Label places where energy is going to waste.

Choose from one of the following situations (or think of one of your own):
- the kitchen
- a teenager's bedroom
- going to school
- the city centre

Ways with words

3 When we talk about conserving energy resources, we mean that we should not waste them. Other people might use the word *conserve* or *conservation*. Here are some people who might use these words.
- A museum keeper: 'My job is to conserve Egyptian mummies.'
- A wildlife ranger: 'I help with nature conservation in the game park.'
- An athlete: 'It's important to conserve some of my strength for the last lap of the race.'

For each person, find another way of saying the same thing, without using the words 'conserve' or 'conservation'.

Make more of maths

4 Jasmine's mother is thinking of buying a car from the second-hand car salesman. She wants to be sure that it will be cheap to run. He shows her three cars.
- The red car needs 7.0 litres of petrol to go 100 km.
- The blue car needs 8.2 litres of petrol to go 100 km.
- The green car needs 7.5 litres of petrol to go 100 km.

  - **a** Put the cars in order, starting with the one which goes furthest on a tank full of petrol.
  - **b** A bus needs 40 litres of fuel to travel 100 km. Does this mean that it uses more energy than a car?

Thinking skills

5 Here are some examples of energy resources:

coal wood running water sunlight wind

  - **a** Add some more to the list.
  - **b** Make a mind map which shows that some energy resources come from living material.
  - **c** Add to your map to show that some energy resources are renewable and others are non-renewable.

Extension question

6 John sees an advert for electric heating. It says that electricity is a 'clean' way of supplying energy.
  - **a** Explain why electricity is a cleaner way of heating a house than coal.
  - **b** Electricity is often generated by burning coal. Explain why this means that electricity is not as 'clean' as the advert suggested.

SAT-STYLE QUESTIONS

1 Moira's house is on a remote island. Her parents have installed solar cells on the roof, and a wind turbine. These generate electricity.

The house also has a diesel generator which generates electricity when the solar cells and wind turbine are not working.

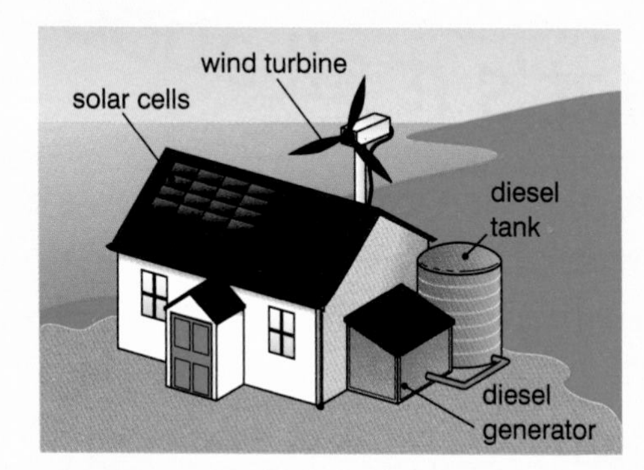

a For each method of generating electricity, choose the energy resource that it uses. (3)

Method	Energy resource
diesel generator	sunlight
solar cells	running water
wind turbine	chemicals
	moving air

b The solar cells do not work at night. Explain why not. (1)

c The wind turbine does not always supply electricity. Explain why not. (1)

d Explain why it is important for Moira's family to have a diesel generator. (1)

2 Class 7Y have been studying energy resources. They have made a list, which their teacher shows on the whiteboard:

geothermal biomass oil coal
moving air tidal running water nuclear
solar natural gas

a From the list, name three fossil fuels. (3)

b From the list, name three renewable energy resources. (3)

3 Class 7Z has been reading about a new tidal power station to be built near their town. The local newspaper published this picture of how it will work: (4)

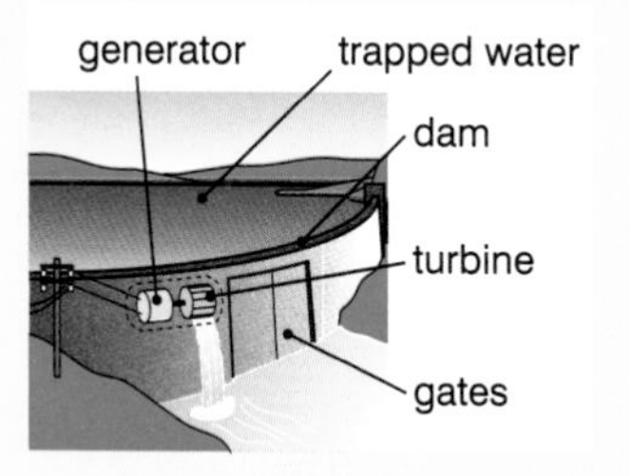

The boxes below show the stages in generating electricity.

A The generator produces electricity.

B The gates shut, trapping the water.

C Water flows out past the turbine, making it turn.

D As the tide comes in, the water level behind the dam rises.

E The turbine turns the generator.

Put the boxes in the correct order. The first one has been done for you.

D				

Key words

Unscramble these:

greeny
lufe
oluje
converse

7J

Electrical circuits

Coming soon

It's made with electric currants.

What's it all about?

We make use of electricity every day – it's one of our most important technologies. It allows us to do all sorts of things at the flick of a switch. Think of computers, music systems, TVs, cars, kitchen appliances. They all depend on electricity.

Electricity can be natural, too. There's electricity in the nerves of our bodies and in every lightning flash.

You should already know quite a lot about simple electric circuits.

In this unit, you will learn to think about electricity and to explain how circuits work.

What do you remember?

You already know about:

- materials that conduct electricity, and materials that don't
- simple electric circuits

1 Which of these materials are good at conducting electricity?

copper plastic steel wood

2 What do we call a material that doesn't conduct electricity?

3 What electrical symbols do you know?

4 If you look at an electric circuit (or a diagram), how can you tell if it will work?

Junior electricians

QUESTIONS

If this lot were in charge of rewiring your home, you'd be in trouble!

- Who has made a complete circuit?
- Who has made a dangerous circuit?
- Does it matter what colour the wires are?

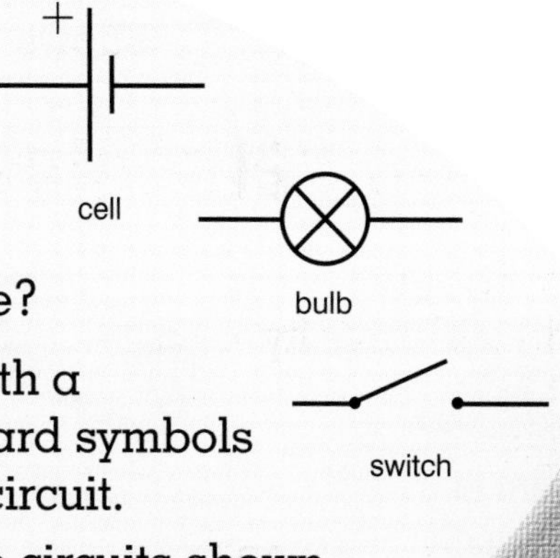

Sometimes it's easier to see a problem with a circuit if you have a diagram of it. Standard symbols represent each of the components in the circuit.

- Can you draw diagrams for each of the circuits shown above? Use your diagrams to explain why some of the bulbs won't light up.

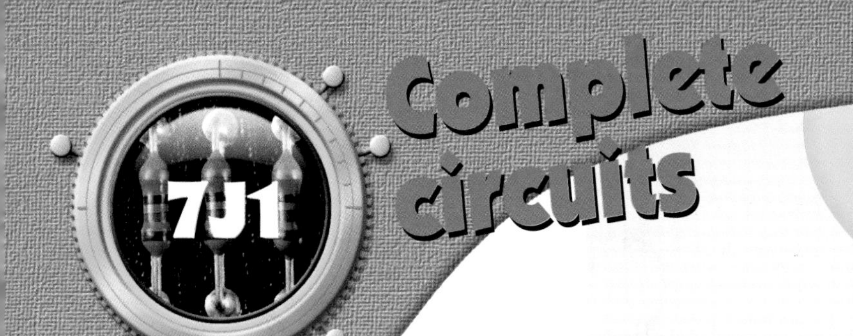

Learn about
- tracing a complete circuit
- circuit symbols
- conductors and insulators

Torches and batteries

It can be annoying if you have a torch that won't work.

- Perhaps chemicals have leaked out of the batteries.
- Perhaps the batteries are 'dead'.
- The metal inside the torch may have gone rusty.
- The bulb may be dead.

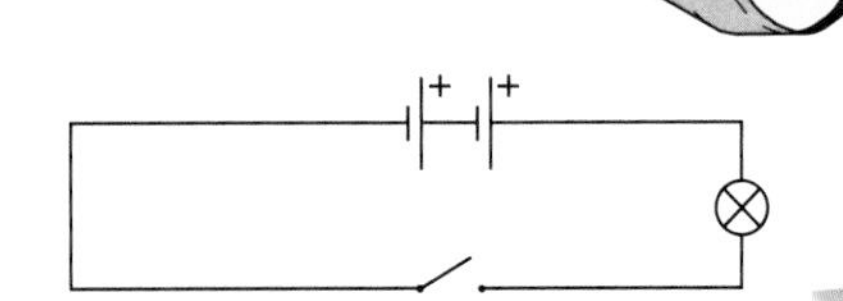

Lots of things can go wrong to stop the electricity from getting to the bulb.

A torch has a battery, a switch and a bulb. There must be a complete circuit of metal from one end of the **battery**, through the switch and the bulb, and back to the other end of the battery.

Q1 Why must the circuit be metal?

Getting the picture

Look at the circuit diagram for the torch. The circuit diagram shows the order in which the components are connected. It also shows that there is a complete circuit.

Use your finger to trace around the complete circuit, starting at the positive (+) end of the **cells**.

Q2 Draw the symbol for a switch. Explain how it shows the way a switch works.

In Technology, you can learn to connect up control circuits and to solder wires.

a cell

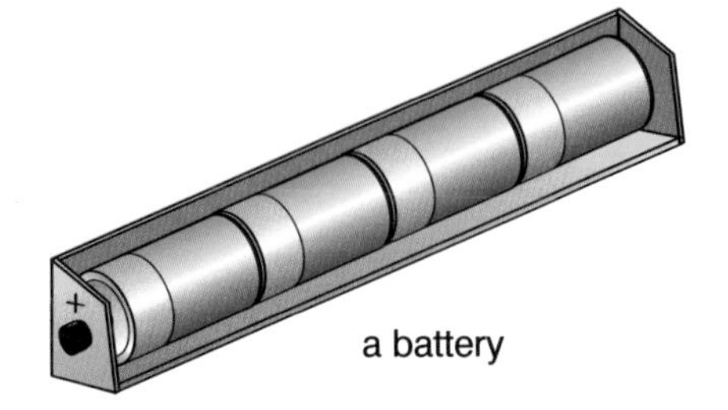

a battery

In science, a single 'battery' is called a cell. Two or more cells together make a battery.

A look at wires

The connecting wires you use at school are covered with plastic. This is **insulation**. Insulation is any material that doesn't conduct electricity. Therefore, if one wire accidentally touches another, electricity can't jump between them.

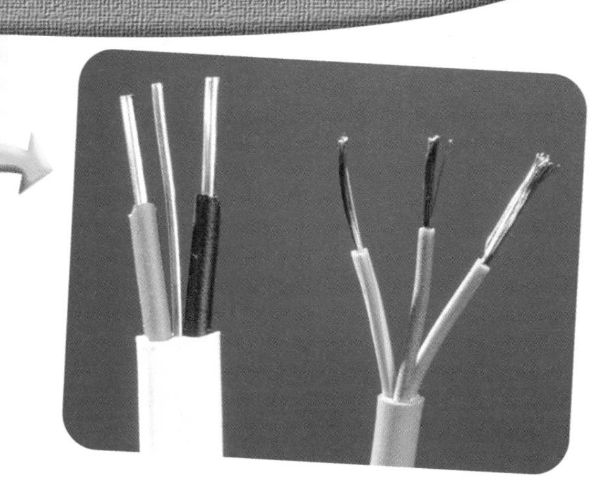

Inside, the wires are made of metal – usually strands of copper, a good conductor. They are quite thick to make it easy for electricity to pass through.

Q3 Explain why you won't get a shock if you touch an insulated wire.

Circuit challenge

- Design and construct a circuit using cells, switches and bulbs.
- Draw a circuit diagram to match your circuit.
- Put your circuit with the others from the class. When your turn comes, display your diagram and see who can match it to the correct circuit.

It worked OK last winter!

SUMMARY QUESTIONS

1 ☆ Copy the circuit diagram for the torch on the opposite page. Label each component with its name.

2 ☆☆ Copy the circuit shown on the right. Add another wire to the diagram to complete the circuit.

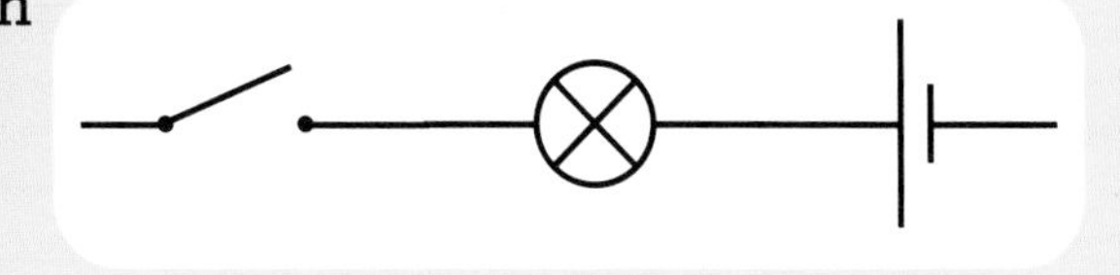

3 ☆☆ After a Science lesson, the pupils complain that some of the bulbs don't work. Design a simple way for the technician to test the bulbs.

4 ☆☆☆ Why is it a good idea to have standard symbols to represent electrical components? Give as many reasons as you can.

Key words
battery
cell
insulation

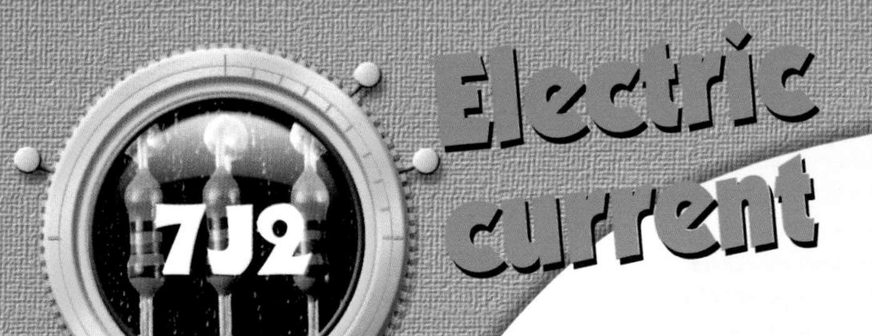

7J2 Electric current

LEARN ABOUT
- electric current
- using an ammeter
- electrical resistance

Flowing around

How do you think about electricity? Is it something that flashes, like lightning? What's going on in the wires when you switch on the light?

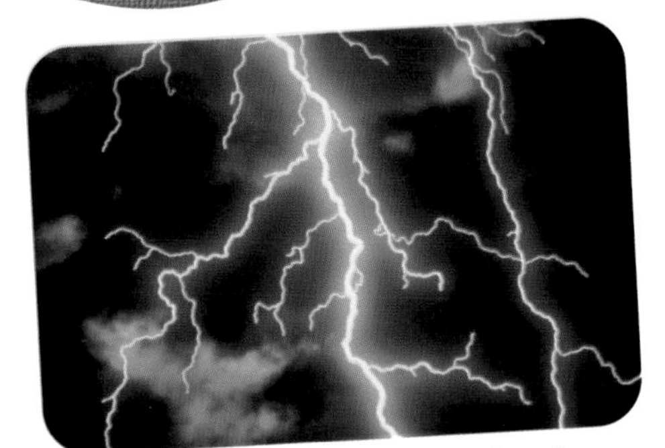

Lightning is a very violent form of electricity

Scientists talk about **electric current**, which flows in the wires. (It's a bit like a current of water, flowing in a river.) So, when you close a switch, the circuit is complete and electric current flows.

Ideas of current

When current flows in this circuit, it makes the bulb light up. It took scientists a long time to work out how the current flows.

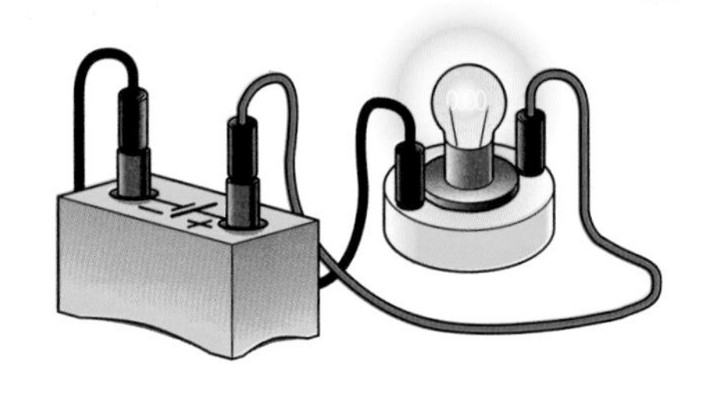

Scientists believe that:
- the current flows from the positive (+) end of the cell –
- through the bulb –
- and back to the negative (−) end of the cell.

The current doesn't get used up along the way.

Q1 The circuit wouldn't work if the black wire was missing. Use the idea of electric current to explain this.

Using an ammeter

An **ammeter** can show us when electric current is flowing in a circuit. We measure electric current in amperes, or amps (symbol A).
This circuit has an ammeter to measure the current flowing *into* the bulb.
- Set up a circuit like this, and then measure the current.
- Now, move the ammeter to a different point in the circuit, so that you can measure the current flowing *out* of the bulb.
- Before you connect up, make a prediction:
  What will the ammeter read now? Will the current be more than before, less than before, or just the same?
- Connect up and find the answer.

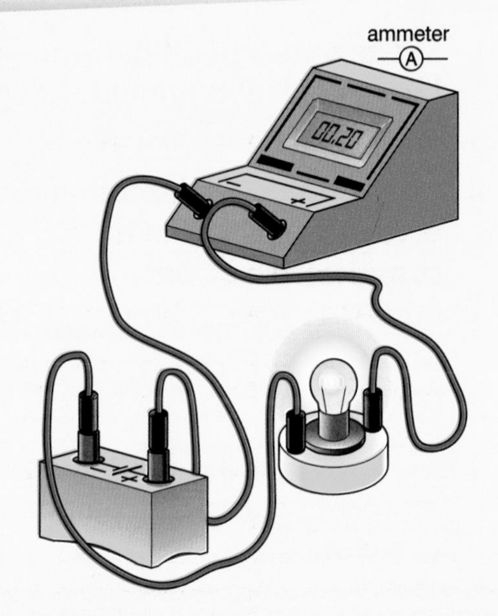

Changing current

What happens if you connect two bulbs in a circuit, one after the other?

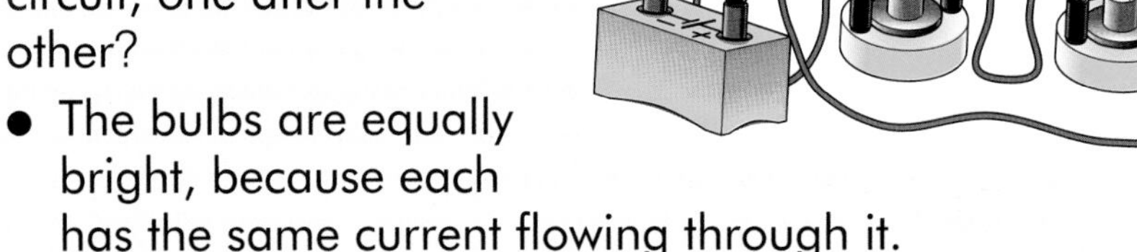

- The bulbs are equally bright, because each has the same current flowing through it.
- The bulbs are dimmer than before, showing that the current is less.

Why is the current less than before?

This is because, with two bulbs, it is harder for current to flow around the circuit. It has to push its way through one bulb, and then through the other. We say that there is more **resistance** in the circuit. Resistance makes it difficult for electric current to flow.

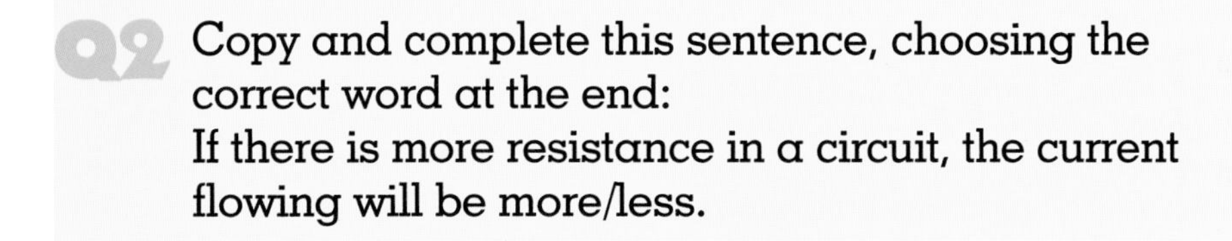

Q2 Copy and complete this sentence, choosing the correct word at the end:
If there is more resistance in a circuit, the current flowing will be more/less.

The volume control on this radio works by changing the resistance in an electric circuit

SUMMARY QUESTIONS

1 ☆ Copy and complete the sentences, using words from the list below.

ammeter   negative   current   positive

Electric . . . flows all the way round a circuit.

Current flows from . . . to . . . .

Current is measured using an . . . .

2 ☆☆ Look at the picture of the electric circuit. Two light bulbs are connected, one after the other, to a cell. They both light up. The ammeter shows that a current of 1 A is flowing into the first bulb.

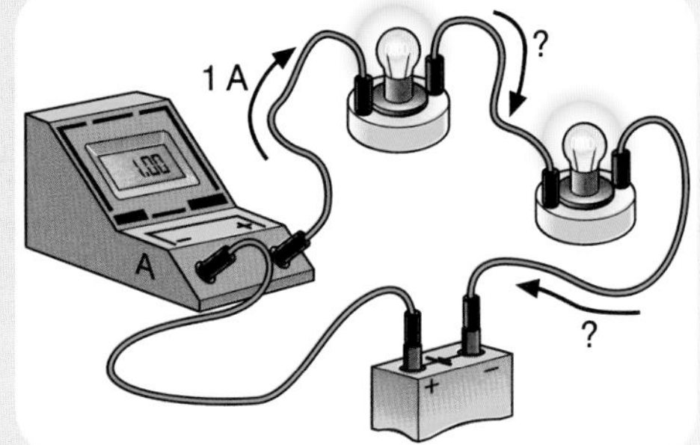

a) How much current flows out of the first bulb and into the second one?

b) How much current flows out of the second bulb?

Key words

ammeter
electric current
resistance

Cells and batteries

Learn about
- voltage
- connecting cells

Choosing the right battery

You need the right battery to make things work. What makes one battery right and another one wrong? The battery must be the right size and shape, of course. But it must also be the correct **voltage**.

Cells and batteries are all marked with their voltages. The letter V stands for 'volt'.

Q1 Look at the batteries in the photo. What are their voltages? What do you notice about these values?

Cell + cell = battery

A 1.5 V 'battery', such as an AA or AAA type, is really a single cell. You may need two or more of these to make something work. For example, a Walkman usually needs 3 V, so you have to put in two 1.5 V cells.

You can probably guess what's inside a 6 V battery. There are four cells, each giving 1.5 V.

Q2 How many cells are needed to make a 4.5 V battery?

Gruesome science

An electric eel can zap you with up to 800 V.

The 'push' of cells

A cell is needed to make a circuit work.
- Try out some different cells and batteries. Connect them, one at a time, in a circuit to light up a bulb.
- See how brightly the bulb lights up.
- Put your observations in a table, and draw a conclusion.

Cell or battery	Voltage	Observation
AA	1.5 V	Bulb not very bright

Take care! Don't connect too many cells to a bulb. It might 'blow'.

The meaning of voltage

The voltage of a cell or battery tells you about the push it can give, to make current flow around a circuit. The greater the voltage, the greater the push. A 3 V battery gives twice as much 'push' as a 1.5 V cell.

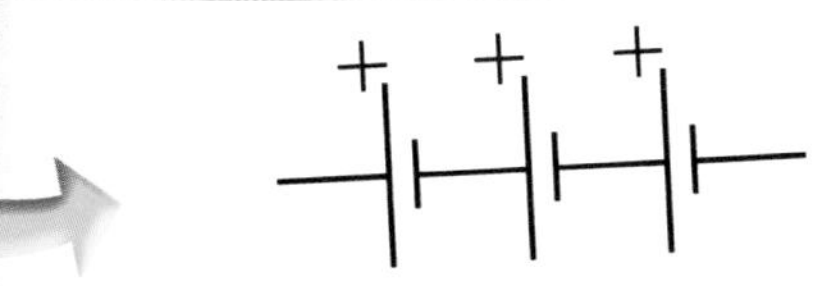

Two cells connected in series give twice the push of a single cell

We can make a battery by connecting together separate cells. The cells must be connected the right way round. They go end to end, with the positive (+) end of one cell connecting to the negative (−) end of the next one, and so on.

If two cells are connected positive to positive (+ to +), their voltages cancel out.

When components in a circuit are connected end to end, we say they are connected **in series**.

Adding up, cancelling out

- Make a circuit with two 1.5 V cells and a bulb. See what happens if you turn one of the batteries round.
- Make a prediction: If you have three 1.5 V cells connected in series, but one is the wrong way round, what will happen? Test your prediction.

SUMMARY QUESTIONS

1 ☆ Copy and complete the sentences, using words from the list below.

series push battery current

The voltage of a cell tells you the . . . it gives to make a . . . flow in the circuit.
To make a . . ., connect cells together in . . . .

2 ☆☆ Look at the diagrams. Work out the voltage provided by each combination of cells.

3 ☆☆☆ Find some electrical items that need more than one cell – for example, a Walkman or battery-powered radio, a cycle lamp, a TV remote control. Look in the battery compartment. Work out which way round the batteries go, and check that they are connected in series.

Key words
battery
in series
voltage

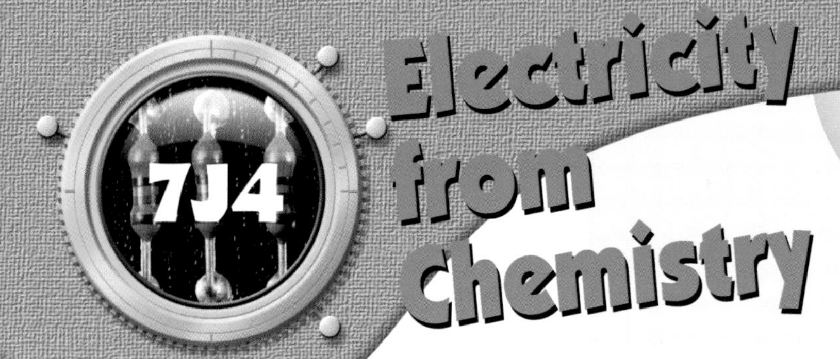

7J4 Electricity from Chemistry

LEARN ABOUT
- energy from a cell
- the difference between current and voltage

Inside a battery

Batteries are convenient, but they soon run out.

Sometimes they leak chemicals which can cause damage.

What's going on?

Inside a cell, there are chemicals. When you connect the cell into a complete circuit, the chemicals start to react with each other. This pushes the current around the circuit.

Disconnect the circuit and the reaction stops.

SAFETY: Never open up a battery. The chemicals inside are hazardous.

Running down

Look at the picture of the circuit. How can we describe what's going on? We have to think about **energy**.

- The chemicals in the cell are a store of energy.
- When the circuit is complete, the energy from the chemicals is transferred to the bulb.
- The energy comes out as light.

Eventually, the chemicals run out, and the battery is 'dead' or 'flat'. Its store of energy has been used up, and the bulb won't shine.

Q1 An AA cell is bigger than an AAA cell. Both have a voltage of 1.5 V. Use the idea of 'energy' to explain why the AA cell will keep a bulb lit for longer. Why is the bulb the same brightness whichever cell is used?

Pedal power

Some bicycle lights work from a dynamo. When the dynamo is working, you have to pedal a little harder. You are providing the energy needed to light the bulb.

Q2 Explain why the lights go out when you stop pedalling.

Current and voltage

It is easy to get confused about the difference between current and voltage. Here are the things to remember.

- Current travels all the way around a circuit.
- Voltage tells you about the push of a cell or battery.
- While a cell is pushing current around a circuit, it is providing energy for the components in the circuit.

Look at the circuit on the opposite page. If you make a break in the circuit, the current can't get through. The light goes out because it is no longer getting energy from the cell.

A large power station supplies electricity at half a million volts. The current leaving it may be over a thousand amps.

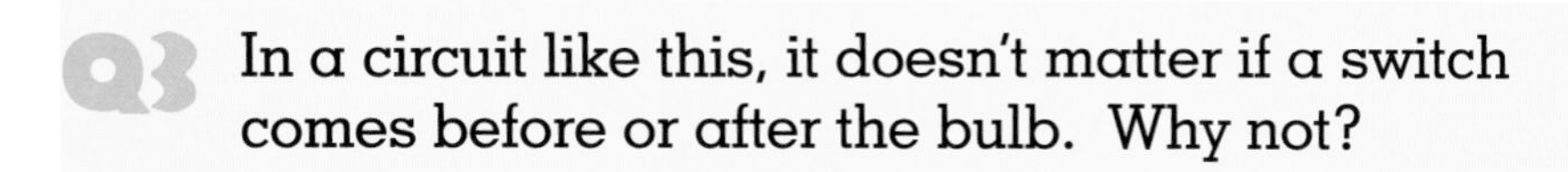

Q3 In a circuit like this, it doesn't matter if a switch comes before or after the bulb. Why not?

Electricity from the mains

Most of our electricity comes from the mains.

Mains electricity is made in power stations. In the UK, most power stations burn either gas or coal. So there is a chemical reaction going on there as well.

When you switch on a light, the power station must burn its **fuel** just a little bit faster.

It's much cheaper to use electricity from the mains than from batteries

SUMMARY QUESTIONS

1 ☆ Copy the table. In each box of the first column, write **current**, **voltage** or **energy**.

	Measured with an ammeter
	Tells you the push of a cell
	Stored in the chemicals of a cell
	Carried by current part way round a circuit
	Flows all the way round a circuit

2 ☆☆ Explain why we use batteries for a torch, rather than plugging it into the mains.

3 ☆☆ Look at the photo of the fan. How can you tell that its batteries are supplying energy?

Key words

energy
fuel
mains electricity

7J5 Series and parallel

LEARN ABOUT
- series and parallel circuits

Switching on and off

If you connect up a circuit with two bulbs in series, together with a switch, you can turn the bulbs on and off. They always go on and off together.

Q1 Explain why the switch turns *both* bulbs on and off.

Your lights at home don't work like this. If you switch one light off, the others stay on. Sometimes two switches control one light! It would be a great nuisance if you had no choice – all the lights in the house on, or all the lights off!

They must be connected up in a different way from in series (end to end). They must be connected **in parallel** (side by side). The picture shows this.

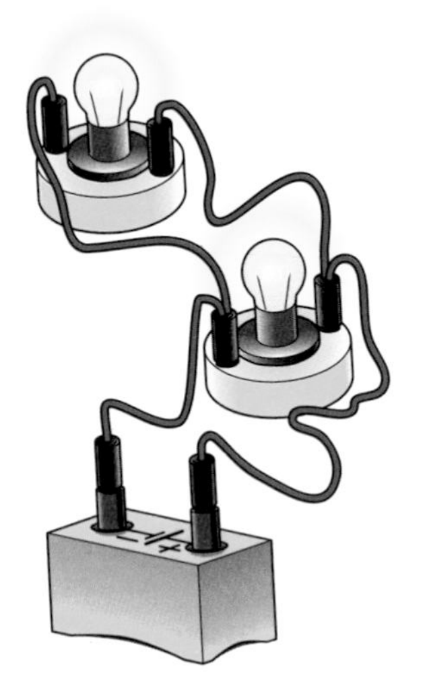

These two bulbs are connected in parallel with each other

Connecting up

- Try setting up these circuits:
  Circuit 1 – a single bulb and a single cell
  Circuit 2 – two bulbs in series with a single cell
  Circuit 3 – two bulbs in parallel with a single cell
- Draw circuit diagrams for each, and note down your observations. (How bright are the bulbs? If there are two, are they both the same brightness?)
  - **a)** In Circuits 2 and 3, what happens if you unscrew one bulb?
  - **b)** In Circuit 3, can you arrange to switch only one bulb off? Can you connect a single switch to switch both bulbs on and off at the same time?

When two wires meet, add a blob to the diagram to show that they are connected together

This picture shows part of the wiring in a house. Red and black wires run all around the house.

Getting wired

The more bulbs you connect in series, the dimmer they get. If you connect bulbs in parallel, they stay just as bright.

Q2 Are your lights at home connected in series or in parallel?

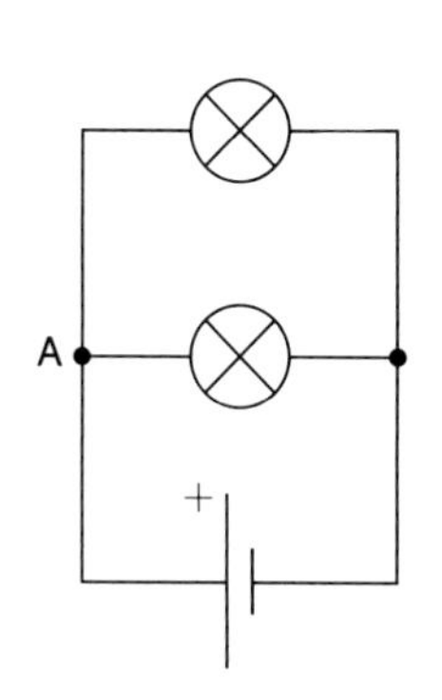

How the current flows

Think about how the current flows around the parallel circuit. Try tracing around the circuit in the picture, starting from the positive (+) end of the cell. What happens when you get to point A? The current can go two ways. So half goes through one bulb, and half goes through the other.

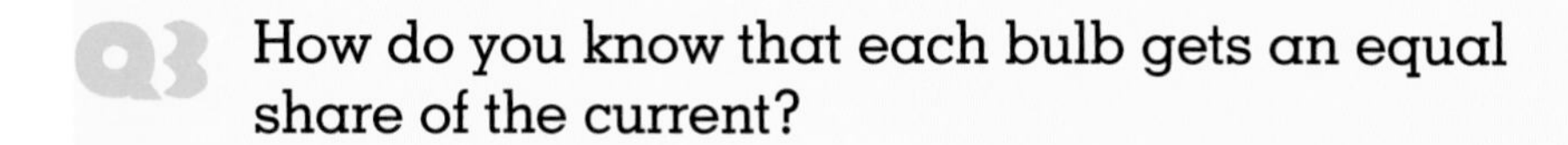

Q3 How do you know that each bulb gets an equal share of the current?

It is easier for current to flow around a parallel circuit than a series circuit, because it has two possible routes to flow along.

SUMMARY QUESTIONS

1 ☆ Copy the table; write **series** or **parallel** in each box in the first column.

	The same current flows through each component.
	There is more than one route around the circuit.
	The current splits as it travels around the circuit.
	Components can be separately controlled.

2 ☆☆☆ Look at the tangled circuit diagrams. Trace the wires round from the positive (+) to the negative (−) end of the cell. For each circuit, answer the following.

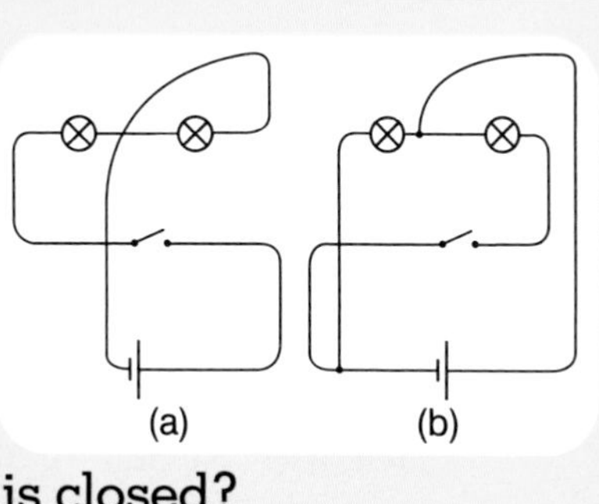

a) Are the bulbs connected in series or parallel?

b) What will happen when the switch is closed?

Make up your own diagram and challenge a friend!

LEARN ABOUT
- electrical safety
- fuses

Danger – high voltage!

Electricity can kill. We know this because people sometimes get killed by lightning, or if they touch the mains electricity. The mains voltage is 230 V. This can push a much bigger current through you than a 1.5 V cell.

Domestic appliances are designed so that you cannot easily get at the high voltages inside.

This hazard symbol warns you of electrical dangers

Fuse protection

There are lots of safety devices used in electrical systems. Here's one – a **fuse**.

A bulb will blow if too big a current flows through it. The circuit in the picture shows how to investigate this. Keep the power supply switched off until the circuit is set up.

The fuse wire is connected in series with the bulb.

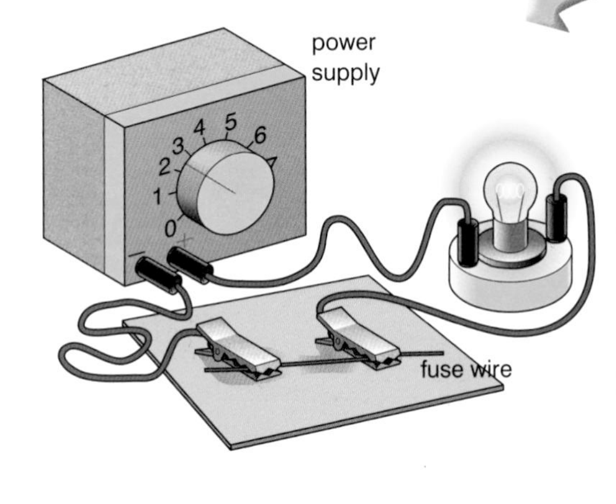

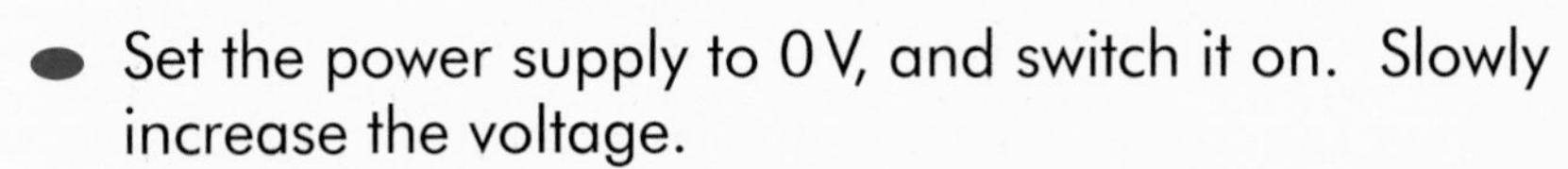

- Set the power supply to 0 V, and switch it on. Slowly increase the voltage.
- Watch the bulb and the fuse wire.

  a) What happens?

Fuse wire is cheap and easy to replace.

  b) Explain how the fuse wire protects the bulb.

There is a small cartridge fuse inside every mains plug. Inside the cartridge is a thin piece of fuse wire.

More safety

Some circuits have **trip switches** instead of fuses. They automatically break the circuit when the current gets too big. They can be reset as soon as the problem has been sorted out. That's much easier than replacing a fuse. Look for the 'RESET' button on the front of a lab power supply.

Q1 Explain why a fuse has to be connected *in series* in a circuit.

Staying alive

It is sensible to have a circuit breaker when using electric tools in the garden. If you accidentally mow through the power cable, you will be protected from getting a shock.

It doesn't take much electric current to kill a person – less than 0.1 A. So we need to be protected from currents flowing through our bodies. Your science lab probably has a **circuit breaker** to protect you.

Suppose you are working in the lab and you accidentally get a shock from a power supply. A small current flows through your body, but almost instantly there is a 'click!' and the power is switched off by the circuit breaker on the wall. You might feel a bit of a shock, but it won't last long enough for you come to any harm.

Q2 You must be careful to avoid touching electrical equipment with wet hands. Why is this?

Gruesome science

Each year in the UK, about forty 11-year-olds die in accidents with electricity. Don't let it be you!

SUMMARY QUESTIONS

1. ☆ Skim through these two pages. Name three electrical safety devices mentioned here.
2. ☆☆ Find out how the teacher resets the circuit breaker in your science lab if it switches off the current.
3. ☆☆ Find out where the 'fuse box' is in your house. Does it have modern trip switches, or old-fashioned wire fuses? (Ask an adult to help you look – but don't touch!)

Key words

circuit breaker
fuse
trip switch

Shocking stuff

Here is another connection between animals and electricity. The horse brake shown in the photo was invented in the 1880s. There is a battery beside the driver. When he pulls a lever, it sends a current through the reins to the horse's mouth. The horse gets a shock and pulls up in a hurry.

The horse brake seems cruel to us today. Animals soon learn to avoid electricity.

An electric fence gives a mild shock to any animal (or person) that touches it. Watch out! To save energy, the fence is only live every few seconds. So you may think a fence is switched off, then get a nasty shock!

If you have a heart attack, your nerves may not be able to control your heart beat properly. An electric shock from a defibrillator may restore its regular rhythm. The paramedic in the photo is giving a patient a shock – you have probably seen scenes like this in TV dramas.

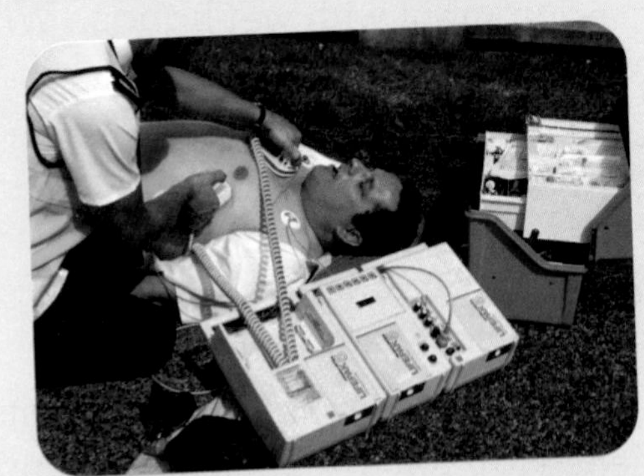

Q1 Why are people nervous about electricity? Do you think that people tend to be too cautious, or not cautious enough?

Nervous about electricity?

Our nerves work by electricity. Tiny currents flow along them, allowing the brain to control our organs. The connection between the body and electricity was first discovered by an Italian scientist called Luigi Galvani. He made simple electric circuits with frogs' legs in them. He saw the legs twitch when he completed the circuit. At first he didn't understand what was going on, but his work soon led to the invention of cells and batteries.

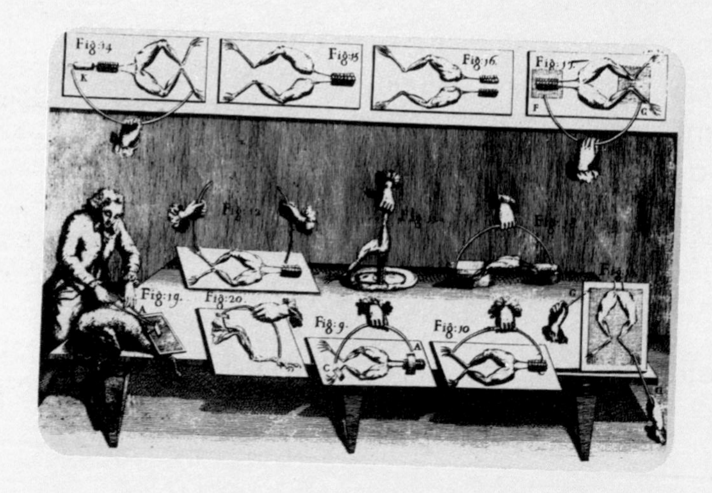

The photo comes from Galvani's book, which was published in 1791. You can see some of the arrangements he tried out. It's a bit surprising to discover that nerves and muscles still work when an animal has died. Many scientists came to believe that it was 'animal electricity' that kept us alive, and that this made the difference between animals and other living things.

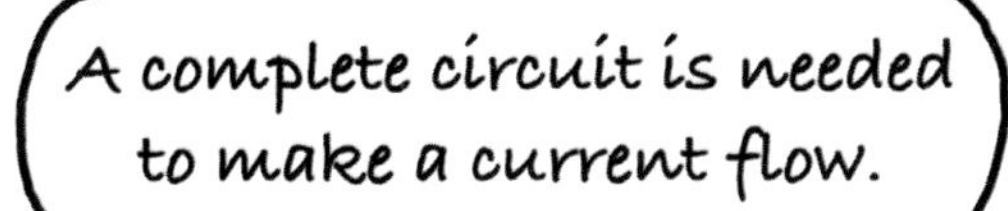

A complete circuit is needed to make a current flow.

The current flows through one component after another in a series circuit... it doesn't get used up.

1·5V
The voltage of a cell tells you how much push it gives a current.

Current is measured with an ammeter.
A

Current divides in a parallel circuit.

The current carries energy from the cell to the components in the circuit.
ENERGY RELEASED
ENERGY STORED

- Electric current needs a complete circuit if it is to flow.
- Current is measured in amps (A) using an ammeter.
- Resistance makes it difficult for a current to flow around a circuit.
- Components can be connected in series (end-to-end) or in parallel (side-by-side).
- The voltage of a cell or battery tells us the push it provides to make current flow.
- Electric current carries energy from the cell or battery to the other components in a circuit.

DANGER! AVOID THESE COMMON ERRORS

It can take a while to understand the difference between voltage and current. **Voltage** is the push. **Current** is what gets pushed around the circuit. Current flows. Voltage doesn't go anywhere.

It may help you to trace the path of current around a circuit. Look for the cells first, and think about the push they give to the current. Then use your finger to follow the path of the current. Remember that, if it's a **series** circuit, the current will simply flow through one component after another. If it's a **parallel** circuit, the current will divide, and then join up again. You may need to use two fingers!

Key words

current
parallel
series
voltage

UNIT REVIEW

REVIEW QUESTIONS
Understanding and applying concepts

1 The picture shows four electrical components.

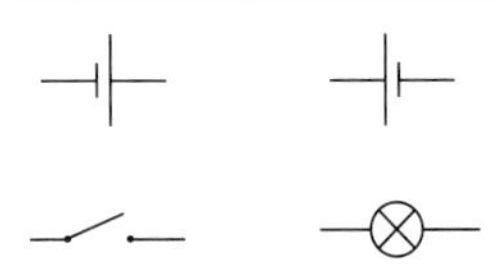

a Make an exact copy of the picture. Add connecting wires to make a circuit so that:
- when the switch is open, the bulb is off.
- when the switch is closed, the bulb is on.

b Make a second copy of the picture. Add connecting wires to make a different circuit. Explain what happens when the switch is open, and when it is closed.

2 What electrical devices are described here?

a This device includes a thin wire that heats up and glows when an electric current flows through it.

b This device has two positions. In one position, it allows an electric current to flow through it.

Ways with words

3 A fuse protects an electric circuit. If a high current flows, we say that the fuse 'blows'. Which of the words in the list best describes what happens to the fuse? Explain your choice.

The fuse … .

explodes **breaks** **melts** **blows up**

Making more of maths

4 Melanie is finding out about electrical resistance. Her teacher gives her several electrical components. Melanie connects each one in turn to a battery and an ammeter. She makes a bar chart to show her results.

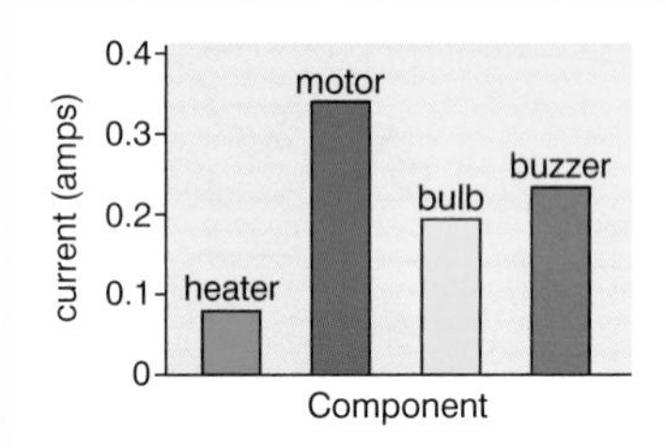

a Which component has the lowest resistance?

b Make a list of the components, starting with the one with the lowest resistance.

Thinking skills

5 We use standard symbols for the components used in electric circuits.

a Draw some standard symbols which could be used in a cookery book. The symbols would represent the different kitchen implements used by cooks.

b Why might these symbols be useful?

Extension questions

6 Jenny is measuring the current flowing through a bulb. She connects it to a single cell and finds that a current of 0.2 A flows.

Her teacher then gives her a second bulb, identical to the first one. Make predictions for the results of the following experiments.

a Jenny connects the second bulb to a single cell.

b Jenny connects the two bulbs in parallel, and then connects them to a single cell.

c Jenny connects the two bulbs in series, and then connects them to a single cell.

In each case, say what current you would expect to flow in the circuit, and why.

SAT-STYLE QUESTIONS

1 Fred sets up the circuit shown. He has three bulbs to test. Each gives a different reading on the ammeter.

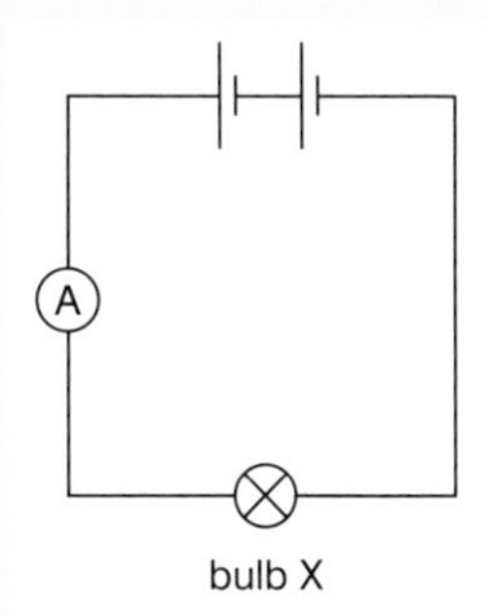

Bulb	Ammeter reading, in amps (A)
X	0.4
Y	0.2
Z	0.6

a Which bulb has the least resistance? (1)
b Which bulb has the greatest resistance? (1)

Fred then sets up a circuit with two bulbs connected to the same battery as before.

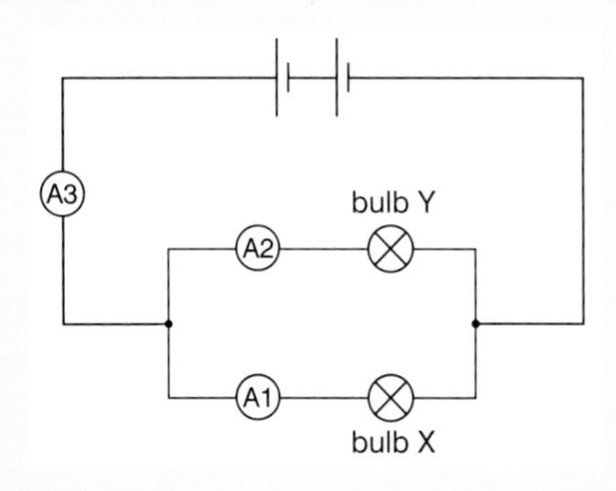

c Are the bulbs connected in series or in parallel? (1)
d Suggest two changes you can make to the circuit to increase the reading on ammeter A1. (2)

2 Omar's teacher gives him an electrical 'black box' to investigate. Omar connects the black box to a cell. He includes an ammeter in the circuit to measure the current flowing through the black box.

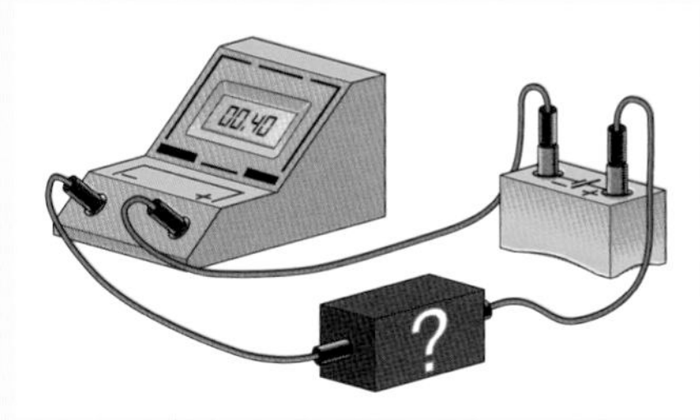

Omar repeats his experiment using more cells and records the current each time. The table shows his results.

Number of cells	Current flowing, in amps (A)
1	0.40
4	1.60
2	0.80
3	1.20

a When the teacher looks at Omar's results, she suggests that he has made his measurements in a strange order. Make a new table, with the results in a sensible order. (2)
b Draw a graph of Omar's results. Draw a line to show the pattern of his results. (2)
c Use your graph to make a prediction: What current would flow if the black box was connected to 5 cells? (1)

3 Jane goes camping. She forgets to turn off her new torch before she goes to sleep. It shines brightly for several hours, then grows dim and fades out.

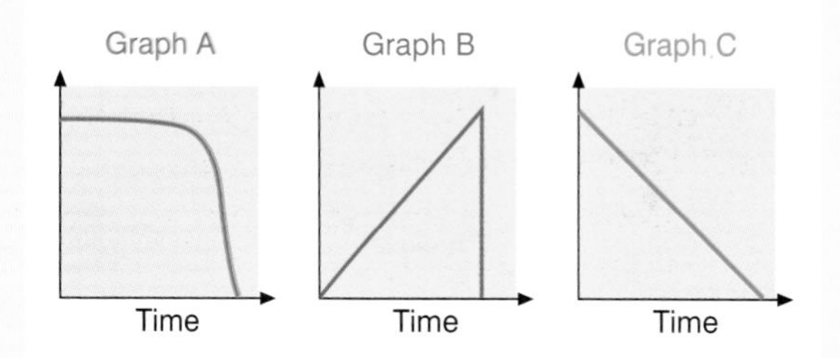

a Which graph represents how the current flowing from the battery changed during the night? (1)
b Which graph represents how the energy stored in the battery changed during the night? (1)

Key words

Unscramble these:
ellc
tybrate
meemrat
galvote
truncer

7K Forces and their effects

Coming soon

What's it all about?

You already know quite a lot about forces. But think back a few years. Before you were born, you floated around in your mother's womb, and you probably didn't have to bother much about forces.

Then you were born. It probably took you about a year to learn to walk, and you must have learned about gravity by falling over. Gradually, you will have learned about different forces and how to make use of them.

Scientists and engineers need to have very clear ideas about forces if they are going to understand the world, or make things such as cars, bridges and boats. In this unit you are going to learn more about the way scientists think about forces.

What do you remember?

You already know about:

- forces such as weight, upthrust and friction
- how to measure forces

1. What force pushes up on us when we are swimming?
   gravity flotation
   upthrust water resistance

2. A crumpled sheet of paper falls more quickly than a flat sheet. Why is this?

3. What force pulls down on us when we are swimming?
   gravity flotation
   upthrust water resistance

4. What unit is used for force?
   kilogram newton
   archimedes forcemeter

5. No-one likes a bumpy road, but roads must have rough surfaces. Why is this?

Fun with forces

QUESTIONS

These students have studied forces, but they seem to have forgotten almost everything they learned.

- Which forces have they named?
- Can you name any forces they have forgotten about?

The students would like their teacher to organise a trip to a fun park on the Moon.

- What would it be like to be on the Moon?

Measuring forces

LEARN ABOUT
- measuring forces
- representing forces

The effects of forces

You stub your toe, you bump your head on a low branch, you trip over the dog. Life can be painful. If you hammer your thumb instead of a nail, the force of the hammer on your thumb hurts a lot.

Forces are not something we can see, but we can see what they do:

Forces can change the shapes of things, and how they move – as these unfortunate drivers discovered!

The racket exerts a force on the ball, and the ball speeds off in a different direction

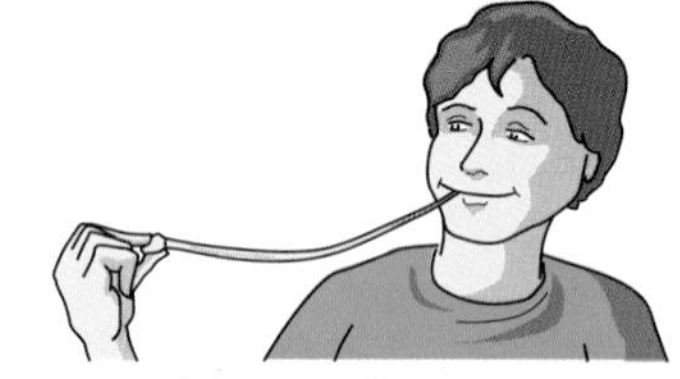

The force with which you stretch the gum makes it change shape

Q1 Give another example of a force changing movement.

Q2 Give another example of a force changing the shape of something.

Measuring forces

Forces are measured using **forcemeters**. A forcemeter has a spring inside it. When a force pushes or pulls the spring, it is squashed or stretched and you can read off the size of the force from the scale.

The meter shows the size of the force in **newtons**. We usually write N for newton.

One newton (1 N) is roughly equal to the weight of an apple

The big push

School labs have bathroom scales which show forces in newtons. Try one out.

- Try pushing on it with your arms and then with your legs. What is the biggest force you can push with?

Picturing forces

The drawings show some force arrows. The label tells us:

what kind of force it is – push, pull etc.	**the pull . . .**
what causes the force	**. . . of the hand . . .**
what the force is acting on	**. . . on the door handle**

Q3 Make a drawing of someone pushing a shopping trolley. Use an arrow to show their pushing force, and label it.

We draw arrows to represent forces. An arrow shows the direction of the force.

Choosing forcemeters

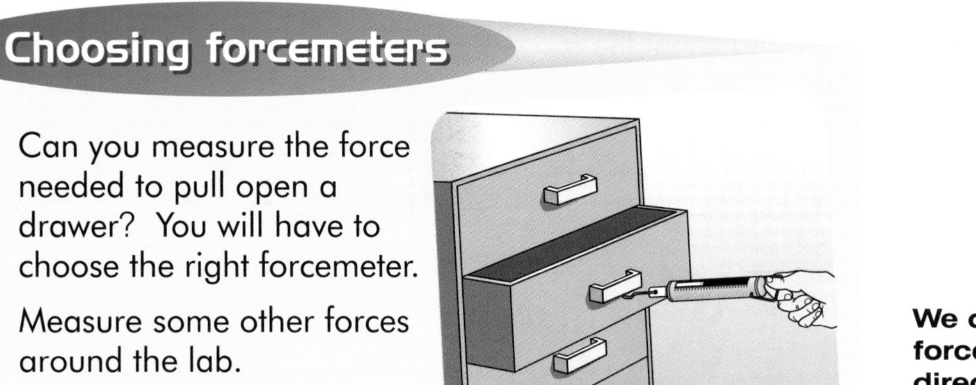

- Can you measure the force needed to pull open a drawer? You will have to choose the right forcemeter.
- Measure some other forces around the lab.

The four engines of a jumbo jet provide a force of 1 million newtons.

SUMMARY QUESTIONS

1 ☆ Copy and complete the sentences, using words from the list below.

newton meter (forcemeter) N arrows newton

Forces are measured using a . . . .

The scientific unit of force is the . . . (symbol . . .).

On diagrams, forces are represented by . . . .

2 ☆☆ Draw a diagram to show a foot kicking a ball. Show the force of the foot on the ball. Remember to label the force arrow correctly.

3 ☆☆ People go to the gym to build up the strength of their muscles. Draw a picture to show how they could use a forcemeter to measure the pulling force of their arm muscles.

Key words

forcemeter
newton (N)

LEARN ABOUT
- weight
- stretching springs

This ornithologist is weighing a small bird before releasing it. It doesn't weigh much – perhaps one-tenth of a newton.

Weight and gravity

You can use a bathroom scales-type newton meter to measure your weight. It might be 500 N. **Weight** is the name we give to the pull of the Earth's **gravity** on us, or on any other object. Weight is a force.

Representing weight

We represent the force of weight by an arrow pointing downwards, towards the centre of the Earth. If there was a hole all the way to the centre of the Earth and you fell into it, you would be pulled all the way to the bottom.

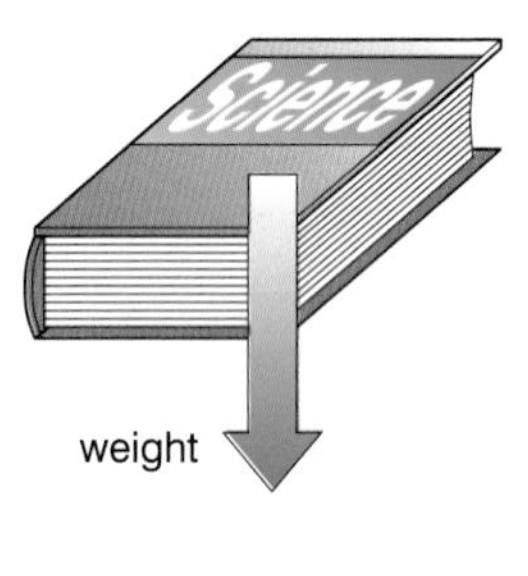

Your weight is a little less if you climb a mountain – because you are further from the centre of the Earth.

Q1 Draw a diagram to show yourself, standing on the ground. Add a force arrow to show your weight. Label the force arrow.

Stretching forces

Most forcemeters have a spring inside them. Even bathroom scales have a spring, a very stiff one. When you stand on the scales, the spring bends and makes the reading on the scale change. The heavier you are, the more the spring bends.

Why is a forcemeter useful in the kitchen?

Investigating a spring

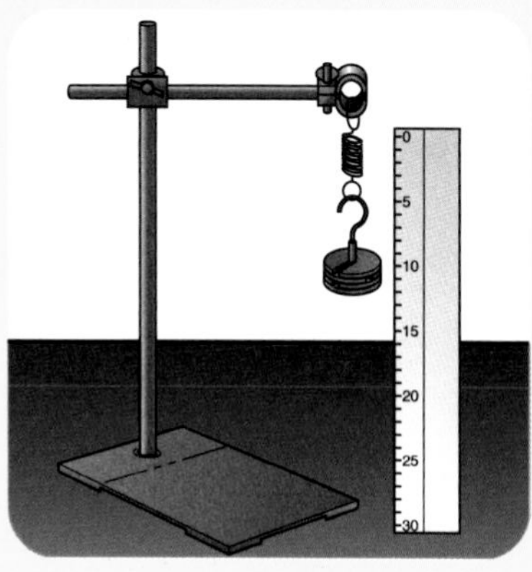

- You can investigate the effect of forces on a spring using the apparatus shown in the picture. The top end of the spring is fixed firmly. When you hang weights on the end, their weight provides the force to stretch the spring.
- Pip and Reese tried this experiment. Here's the pattern they found in their results.

  'Every extra 5 N made it stretch by another 1 cm.'

  Do you find the same pattern in your results?

SUMMARY QUESTIONS

1 ☆ Copy and complete the sentence, using words from the list below.

pull weight gravity

The . . . of an object is the . . . of the Earth's . . . on it.

2 ☆ What units do we measure weight in? Why do we use these units?

3 ☆☆ We use an arrow to represent the weight of an object. Which direction does the arrow point in?

7K3 Mass and weight

LEARN ABOUT
- the difference between mass and weight

Getting started

When you buy fruit and vegetables at the market or supermarket, the price is given 'per kilogram'. At the checkout, the assistant weighs them to find out how many kilograms you have bought – the scales measure the mass of your shopping.

The **mass** of an object is measured in **kilograms** (kg).

The more kilograms you buy, the greater the mass you will have to carry home.

Q1 It is much easier to carry a bag of lettuces than a bag of potatoes. Use the idea of 'mass' to explain why.

Mass tells us about how much **matter** something is made of. A lettuce has less mass than a potato because there is a lot of air between the leaves. The air has very little mass.

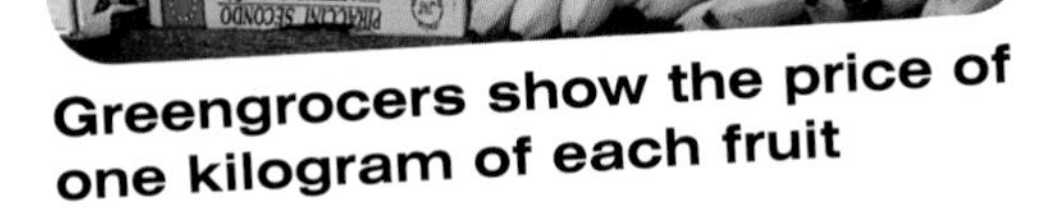

Greengrocers show the price of one kilogram of each fruit

Getting weighed

The idea of mass sounds a lot like the idea of weight. We talk about weighing ourselves, but the answer comes out in kilograms, so we have really found our mass. Try to remember these important differences.

- Mass in kilograms tells us how much matter something is made of.
- Weight in newtons is a force – it tells us about the pull of gravity on something.

My mum says I was a ten pound baby...
...I'm sure I cost more than that!
PHOTOS

Judging weight

We use force meters to measure weight, because weight is a force. Try this.

- Hold an object in each hand. Can you tell which one has the greater weight?
- Check your answer by weighing the objects with a forcemeter. Were you correct?

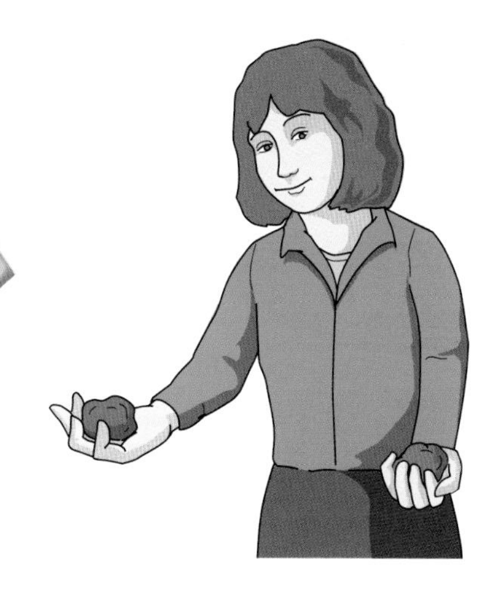

To the Moon

If you went to the Moon, you would be able to jump much higher and throw a ball much further. The Moon's gravity is much less than the Earth's gravity, and so the weight of everything is less – about one-sixth of its weight on Earth.

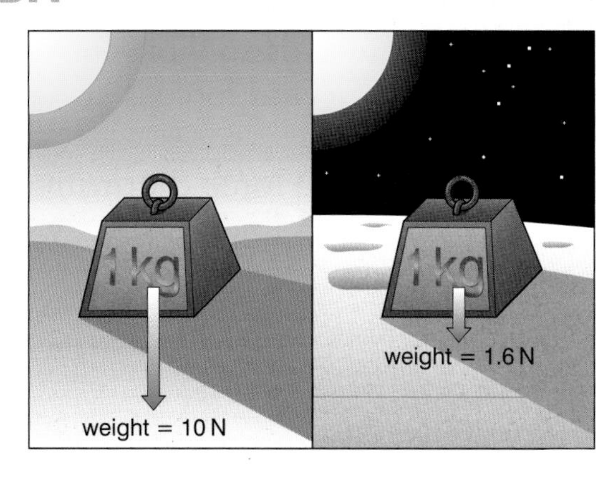

A 1 kg mass has a weight of 10 N on Earth, but only 1.6 N on the Moon

CHALLENGE

People have strange ideas about gravity on the Moon. Try emailing these questions to three or four people, including at least one adult.

- What would you notice if you dropped a stone on the Moon?
- Why is the Moon's gravity different from the Earth's?

Does everyone agrree?

SUMMARY QUESTIONS

1 ☆ Copy and complete the table:

mass	The amount of . . . in an object	measured in . . . (kg)
. . .	The pull of the Earth's . . . on an object	measured in . . . (. . .)

2 ☆ A box has a weight of 1000 N on Earth. Its mass is 100 kg. Draw a diagram of the box, and add a labelled arrow to represent its weight. Label the box with its mass.

3 ☆☆ Jupiter is a giant planet, so its gravity is much stronger than Earth's. How would your weight change if you went to Jupiter?

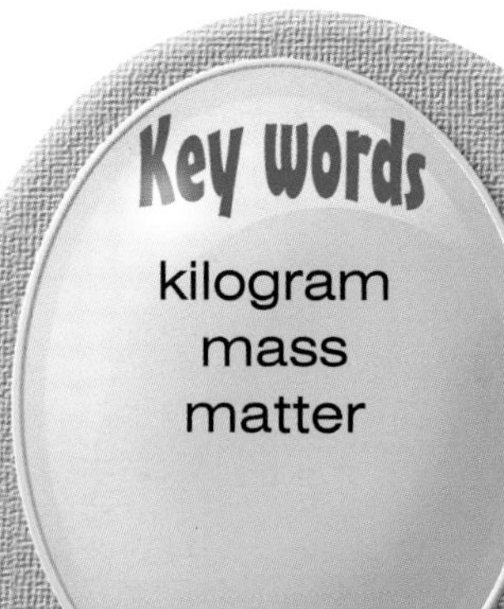

Floating and sinking

Learn about
- forces in water
- balanced forces
- measuring and calculating density

Going under

At the swimming pool, a small child is nervously holding on to the edge. He's sure that he will sink to the bottom if he lets go. He thinks that the deeper the water, the more easily he will sink.

Have you ever tried to sit on the bottom of the pool? It's very difficult to stay completely under water. The water pushes you upwards so that you quickly return to the surface. This upward force is called **upthrust**.

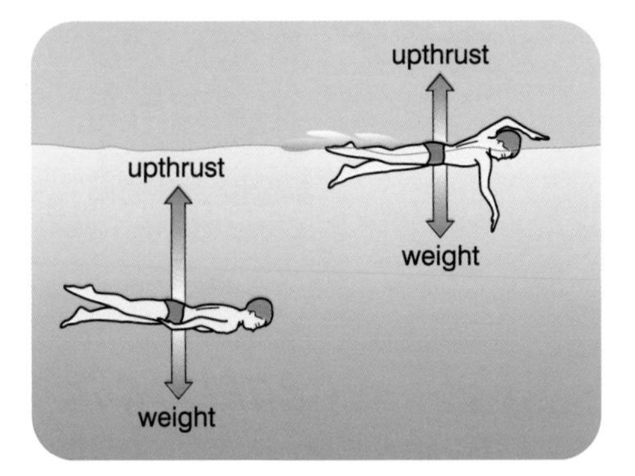

When you are under water, the upthrust pushes you back towards the surface. When you float, the upthrust of the water cancels out your weight.

Q1 Look at the force diagrams of the swimmer. When he is under water, which force is bigger?

Staying still

If you sit on a chair, there are two forces acting on you.

- Your weight pulls you downwards.
- The chair pushes upwards.

These two forces are the same size, so they cancel each other out. We say that the forces are **balanced**.

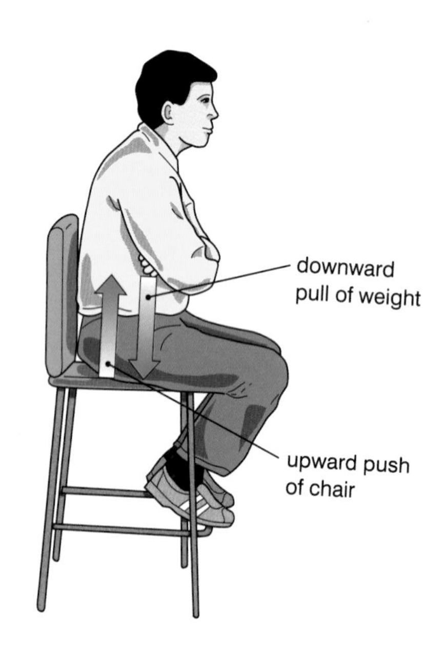

Q2 Draw a diagram to show the forces which act on a ball lying on the ground. What extra, unbalanced force could make the ball move? Add it to your diagram.

Why do we float?

Some things float in water, but others sink. What's the difference? It depends what material things are made of. 'Heavy' materials sink. 'Light' materials float.

A scientist can predict if a material will float by finding out its **density**. A dense material has a lot of mass squashed into a small volume. If its density is greater than the density of water, it will sink.

Fortunately, people are a little less dense than water, so we float. The upthrust of the water is enough to balance our weight.

Comparing densities

You will be given some samples of different materials. Your task is to put them in order, from most dense ('heaviest') to least dense ('lightest').

- Design an experiment to do this.
- Do you get the same answer as everyone else?

Gruesome science

A man who survived the wreck of the *Titanic* fell into a puddle, knocked himself unconscious and was drowned.

SUMMARY QUESTIONS

1 ☆ A duck floats on the surface of a pond. Draw a diagram to show the two forces which are acting on it. Are the forces balanced or unbalanced?

2 ☆☆ The wood from some trees will float in water. The wood from other trees sinks. What does this tell you about these different types of wood?

3 ☆☆☆ A helium balloon will float in the air. There are two ways to explain this.
a) Use the idea of density to give one explanation.
b) Use the idea of forces to give another explanation.

Key words

balanced forces
density
upthrust

Learn about

- friction, drag and air resistance
- lubrication

Smooth talking

Friction can be a nuisance, but it can also be enormously useful to us.

- Suppose you want to push a heavy object along the ground. The force of friction makes this difficult.
- Imagine walking on an ice rink. It's almost impossible because there's hardly any friction.

Opposing friction

Friction appears whenever one surface tries to slide over another. To reduce friction, we can add a **lubricant**. That's why car owners put oil into their car engines from time to time, so that the metal parts can slide smoothly over each other.

Q1 Look at the picture of the children on the slide. What lubricant is at work here? What would it be like without this lubricant?

Investigating friction

Here's one way to find out about friction.

- Place a block of wood on one end of a plank. Gradually raise that end of the plank. Eventually, the block will slide down to the other end. With a lot of friction, you will have to lift the plank a long way.

- Use this idea, or one of your own, to plan an investigation into friction. Try to answer this question: What factors affect the amount of friction between two surfaces?

Submarines have an outer coating designed to imitate the skin of dolphins, which gives as little drag as possible.

Opposing motion

The drawings show some ways we make use of friction.

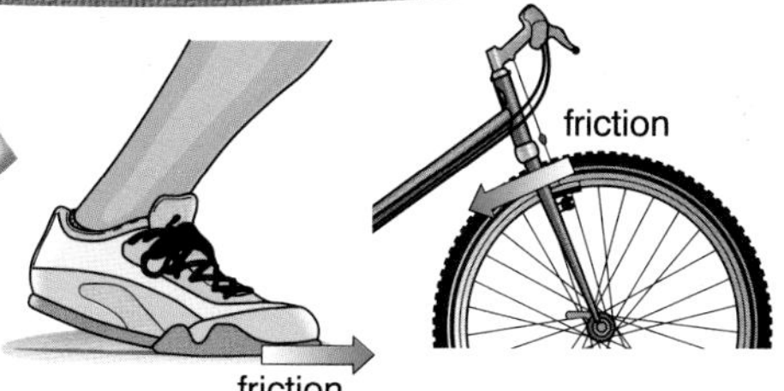

Friction is a force that opposes motion. Like any force, we can use an arrow to represent it on a diagram. You have to think: *In which direction is the object moving, or trying to move?* Then the arrow points in the opposite direction.

Movement in water and air

Friction isn't the only force that opposes motion. Ships and submarines are slowed down by the force of **drag** as they move through the water.

Sharks have evolved to have a shape which cuts quickly through the water, with as little drag as possible

There is drag in air, too, but it's called **air resistance**. There is less drag in air than in water. That's why it's possible to have supersonic aircraft but not supersonic ships – or fish!

Engineers have to understand friction. They design boats to be a good shape for cutting through the water. They give a fast car a streamlined shape.

Q2 Sketch the outline shapes of two cars. Give one a much more streamlined shape than the other.

Parachutists make good use of air resistance to slow them down as they fall

SUMMARY QUESTIONS

1 ☆ Copy and complete these sentences, using words from the list below.

lubricant opposes surfaces

Friction is a force which . . . motion when two . . . try to slide over one another.

Friction can be reduced by using a . . . .

2 ☆☆ Draw a diagram to show a child moving down a slide. Using one colour, draw a labelled arrow to show the direction of their movement. Using a different colour, draw a second arrow to show the force of friction acting on the child.

3 ☆☆ Describe how divers give themselves a streamlined shape, so that they move easily through the air and water.

Key words

air resistance
drag
friction
lubricant

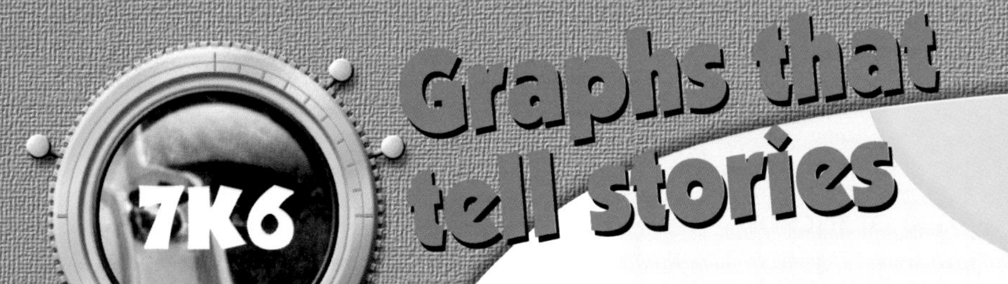

7K6 Graphs that tell stories

LEARN ABOUT
- speed and its units
- distance–time graphs

Top speed

Perhaps you can run at 6 metres per second (6 m/s). On motorways in the UK, the highest permitted speed is 70 mph (112 km/h), although many people drive faster than that. A jumbo jet travels at about 1000 km/h.

Drivers tend to slow down when they see traffic police at work

We use different units for speed, depending on the units we are using for measuring distance and time.

Quantity	Units
Distance travelled	metres, kilometres, miles
Time taken	seconds, hours
Speed	m/s, km/h, mph

Q1 Top sprinters can run 100 m in 10 s. How far do they run in each second?

Going up

The graph shows someone's journey in a lift to the top of a tall building. Make sure you read the labels on the axes of the graph.

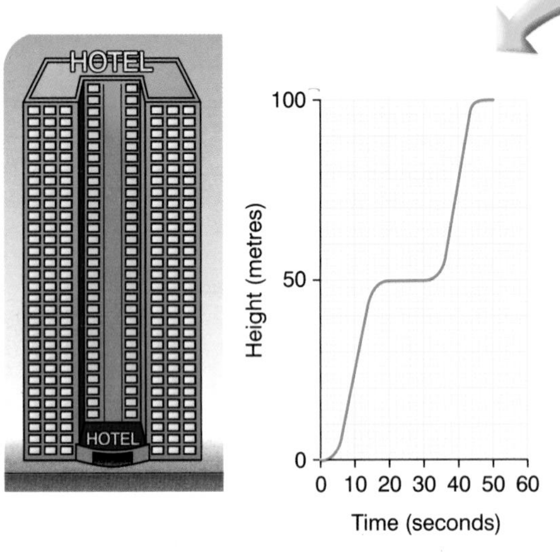

From the graph, we can find out various things about the ride in the lift.

- The building was 100 m high.
- The complete journey took 50 s.

The lift started off slowly, then rose at a steady speed for a while. It slowed down as it came to a halt at the top.

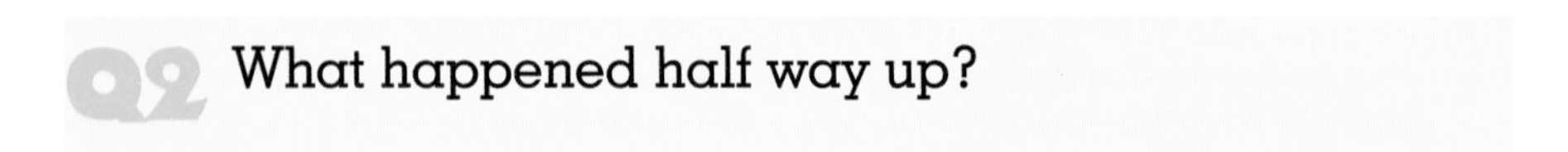

Q2 What happened half way up?

Along the road

The graph shows this journey:

'I walked along the road, and waited at the bus stop. The bus took me all the way to school without stopping.'

Distance

Time

You can see the three parts of the journey on the graph. The line slopes up when I was walking. It is horizontal when I was at the bus stop, and not moving. Then it slopes up again; it is steeper, because the bus was going faster than I could walk.

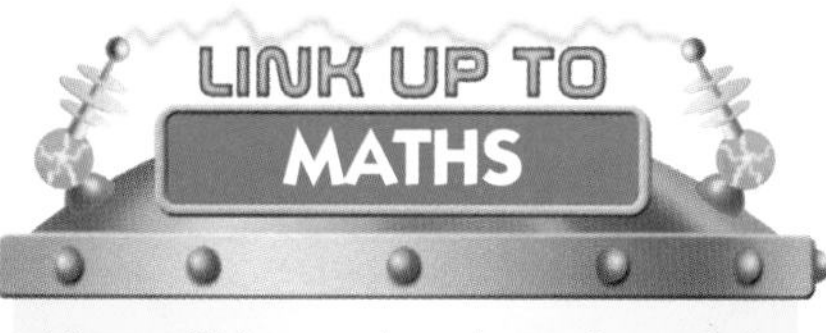

You will have developed useful skills for interpreting and drawing graphs in your studies of maths.

Q2 Draw another graph to show my return journey: I waited at the bus stop; then I travelled on the bus; then I walked home.

Graph challenge

- On a piece of paper, draw a **distance–time graph** for a journey. It doesn't need to have accurate figures on it.
- Next, in a small group, put all your graphs on display.
- Then take it in turns to describe your journeys. Everyone has to guess which is the correct graph.

SUMMARY QUESTIONS

1 ☆ Which of the following is NOT a unit of speed?

m/s mph km/h s/km

2 ☆☆ Look at the graph above. Which section shows the bus ride? How can you tell the bus didn't stop?

3 ☆☆☆ Draw a graph to represent this journey:

'As I was walking to school, I suddenly realised I was going to be late, so I ran all the rest of the way. I had to sit down on the steps when I got there.'

Key words

distance–time graph

speed

 SCIENTIFIC PEOPLE 

Archimedes and the king's crown

You probably know the story of Archimedes in his bath. King Hiero ordered a jeweller to make him a new gold crown, in the shape of a wreath of leaves. He suspected that the jeweller had cheated him by mixing silver with the gold. Could Archimedes find a way of checking the crown without damaging it?

Archimedes was in his bath when he thought of the solution. When you get into the bath, the water level rises. This is because your body displaces some of the water. Archimedes, seeing how he could put this to use, leapt from the bath and ran down the street shouting 'Eureka!' (This is Greek for 'I have it!')

When he was dressed again, he took a pure gold crown that weighed the same as the crown from the jeweller, together with a bowl of water. He submerged each crown in turn in the water. The water level went up more for the jeweller's crown, showing that the crown had a bigger volume.

Archimedes knew that silver is less dense than gold. He realised that his experiment showed that the king's new crown was indeed a cheat, and the jeweller was punished.

Q1 Why was it important to check the new crown?

Q2 Why was it important to check it without damaging it?

Q3 Which takes up more space, 1 kg of silver or 1 kg of gold?

DANGER! AVOID THESE COMMON ERRORS

In this unit, you have learned about some forces – **weight**, **upthrust**, **friction** and so on. You know that forces can do two things.

- They can change the way something is moving.
- They can change the shape of something.

If something isn't moving, the forces on it must be balanced. But because we can't see the forces, we may think that there aren't any forces at work at all. It is important to know that there may be forces there, but their effects are cancelling each other out.

Floating in water is an example. There are two forces at work – weight and upthrust. They are equal in size but they are pushing in opposite directions.

Standing on a frozen pond is another example. Your weight pulls you down, and the ice pushes up on you. If the ice isn't strong enough to balance your weight, you will fall through it.

Key words

friction
upthrust
weight

UNIT REVIEW

REVIEW QUESTIONS

Understanding and applying concepts

1 The pictures show an experimental rocket and the forces acting on it. It is designed to carry a heavy load high into the sky.

A Standing on the ground

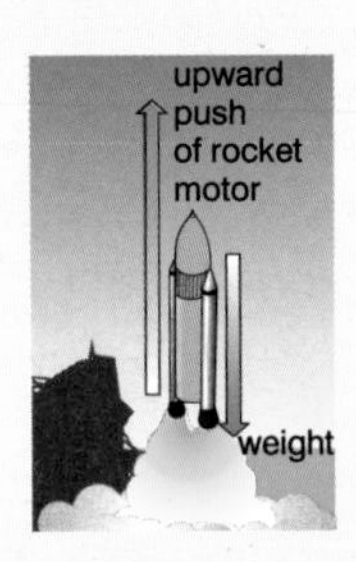

B Just above the ground

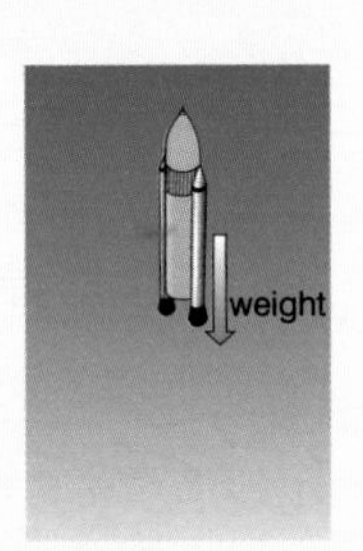

C High above the ground, all fuel burned

- **a** Look at picture A. Are the forces on the rocket balanced? Will it begin to move?
- **b** Look at Picture B. Use the idea of forces to explain why the rocket rises in the air.
- **c** In Picture C, the rocket has burned all its fuel. Use the idea of forces to predict what will happen next.
- **d** Suggest two ways to alter the rocket to make it go higher.

2 Joel accompanies his grandmother to the post office. They walk slowly along the road and then sit on a bench for a few minutes. Then they walk the rest of the way. They have to stand in a queue before they are served.

Draw a graph of distance against time to represent this journey.

Ways with words

3 The newton is the unit of force. It is named after Isaac Newton. Here are two more units, named after scientists.

joule watt

The joule was named after James Prescott Joule. The watt was named after James Watt. Find out about these scientists. What did they study? When did they do their work? What quantities are measured in these units?

Making more of maths

4 Pat and Paul are investigating mass and weight. They have three objects marked with their masses. Pat weighs each one and shouts out the results. Paul writes down the results on a scrap of paper.

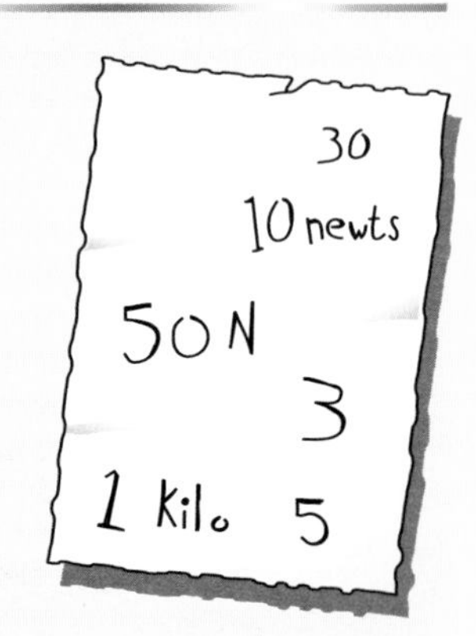

As you can see, Paul didn't do a very good job of this. Can you rescue him by putting the results in a table, with a heading at the top of each column?

Thinking skills

5 Jane picks up two objects, A and B. She weighs them to find their masses. This is what she notices.

- A and B have the same mass.
- B has a smaller volume than A.

Which is more dense, A or B?

Extension question

6 There is friction when one rough surface tries to slide over another. Imagine that you have a very powerful microscope and that you can look at the two surfaces.

- **a** Draw a picture of what you think you would see.
- **b** A lubricant like oil or water can help to reduce friction. Draw another picture to show your idea of how this works.

SAT-STYLE QUESTIONS

1 In an experiment to investigate how a spring stretches, Jay and Kay hang weights on the end of the spring and measure its length.

The table shows their results.

Weight, in newtons (N)	Length of spring, in mm	Increase in length, in mm
0	40	0
1	46	6
2	52	12
3	58	
4	64	
5	70	

a Jay says, 'Every time we increase the weight, the spring will get longer.' Study Jay's prediction, and look at the table of results. Was Jay's prediction supported by the results? (1)

b Copy the table and complete the final column. (1)

c Draw a graph to show the results. (2)

d Use your graph, or the table of results, to find out how much the spring stretched for every newton of load. (1)

e Kay says, 'If we double the weight on the spring, it will get twice as long.' Kay tried to make a better prediction than Jay. But what she said is not quite right. Write down a better conclusion, based on the results they obtained. (1)

2 Emil is coming down a water slide.

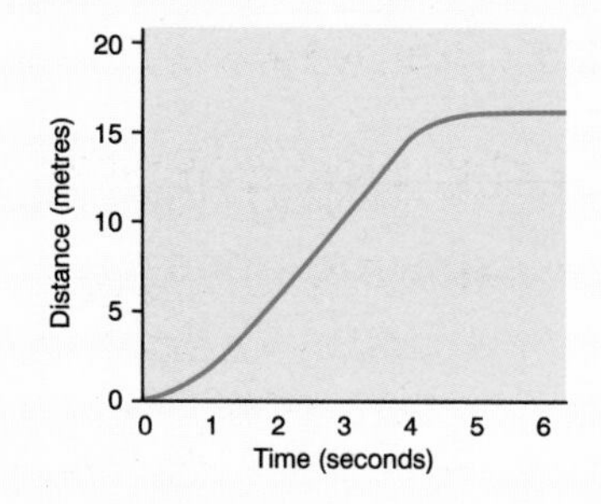

a Name two forces acting on Emil when he is part way down the slide. (2)

b The graph shows Emil's journey down the slide. Describe his motion between 1 and 4 seconds. (1)

c Describe his motion between 5 and 6 seconds. (1)

3 Emma carries out an experiment to investigate floating and sinking. She has a wooden ruler. She weighs a lump of plasticine and attaches it to one end of the ruler. Then she floats the ruler in water and records the length of the ruler sticking out of the water.

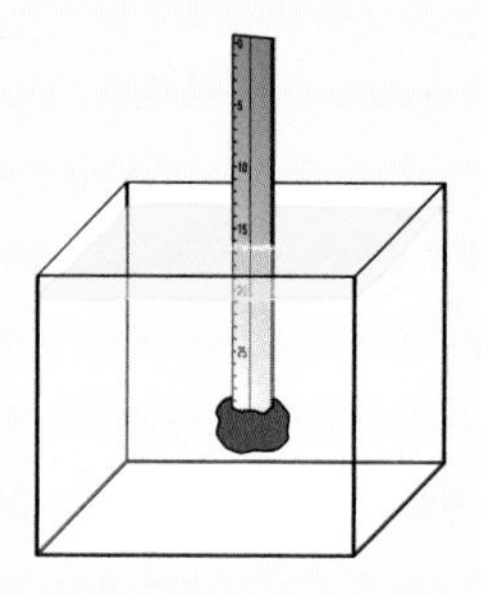

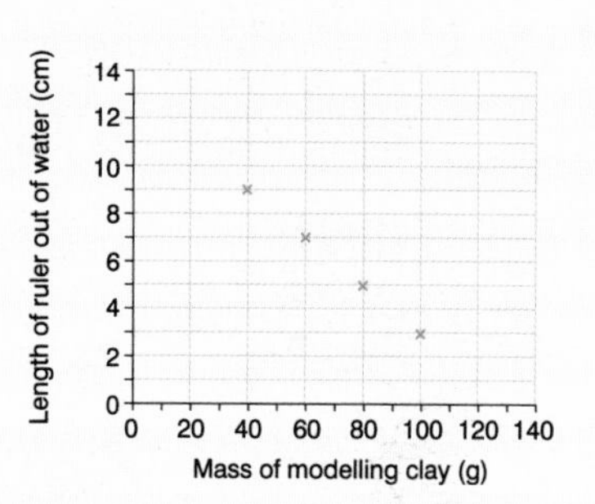

Study the graph of Emma's results.

a Use the graph to predict: what mass of plasticine is needed to make the ruler sink? (1)

b If there was no plasticine, what length of the ruler would stick out of the water? (1)

c Put these materials in order, starting with the most dense: (3)

water wood plasticine

d Emma says, 'As the ruler gets heavier, the upthrust of the water gets less, so eventually the ruler sinks.' Explain why Emma is wrong. (2)

Unscramble these:
nonwet
gard
deeps
viagryt

7L

The solar system – and beyond

COMING SOON

What's it all about?

Space – it's all around us. And it's vast. You could fly off at the speed of light and, even in a million lifetimes, you wouldn't reach the end of it.

Today we know that the Earth and the other planets travel around our star, the Sun. The solar system is just one tiny bit of an ordinary galaxy.

In the Middle Ages, people thought that the Earth was at the centre of everything, with the Sun, Moon and planets travelling around us.

In this unit, you will learn about the solar system, and a little of what lies beyond.

What do you remember?

You already know about:

- the Earth, Sun and Moon
- how we see things
- how shadows are formed

1 Which is biggest?
Earth Moon Jupiter Sun

2 Which is smallest?
Earth Moon Jupiter Sun

3 How long does it take the Earth to spin around once?
1 hour 1 day 1 month 1 year

4 How long does it take the Earth to orbit the Sun once?
1 hour 1 day 1 month 1 year

Observing and explaining

The Moon is just the same size as the Sun.
We only see the Moon at night.

It's dark at night because the Earth has gone behind the Sun.
And in Australia, it's dark in the daytime.

The planets look just like stars...
...but when I look through my telescope, I can see that they are much bigger than stars.

Stars only look smaller than planets because they are a long way away.

QUESTIONS

Some people have funny ideas about space!
In science, we **observe** things and we try to **explain** them.
Sometimes our observations mislead us, and we come up with the wrong explanations.
Look at the amateur astronomers in the cartoon. Some of their observations are incorrect, and some of their explanations are really terrible.

- Can you find anyone to agree with?
- Can you put the others right?

7L1 Light from the Sun

LEARN ABOUT
- the luminous Sun

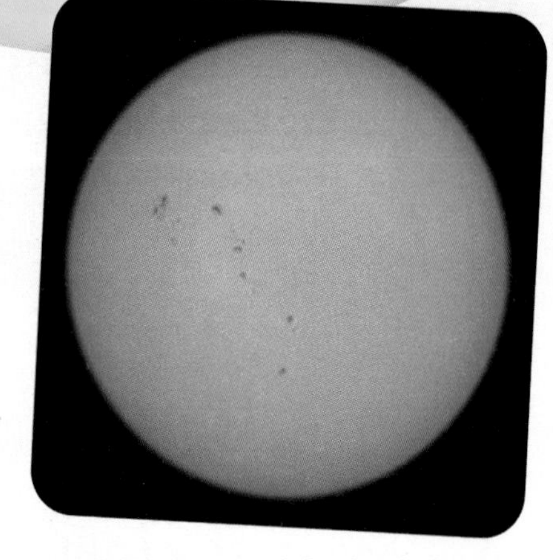

The Sun

The Sun is our **star**. We get heat and light from the Sun. This keeps the Earth at a comfortable temperature, suitable for life.

The Sun is a giant sphere of hot gases. It is a **luminous** object, and without it, the Earth would be a cold, dark lump of rock, drifting in space.

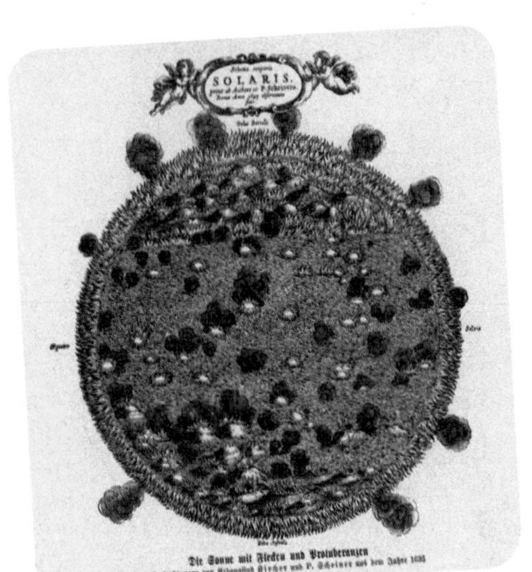

In Victorian times, scientists had different ideas about the Sun. Some thought it might be made of burning coal. Others thought it might be covered with volcanoes, as shown in this picture.

Day and night

When our side of the Earth faces the Sun, it's daytime, and it gets hot. When the Earth turns so that we are facing away from the Sun, it's night-time. Then the temperature drops.

It's fortunate for us that the Earth spins on its axis. If it didn't, the side facing the Sun would be very, very hot. The other side would be dark and freezing cold.

Q1 In which direction does the Sun rise? In which direction does it set?

Q2 Your answers will help you to say: In which direction does the Earth spin?

Seeing the Sun

It is very dangerous to look directly at the Sun. You could permanently damage your eyes – or even blind yourself.

The picture shows a safe way to look at the Sun. Your teacher will set this up for you. If you are lucky, you may be able to see sun spots – cool patches on the Sun's surface.

How could you discover whether the Sun spins around?

SAFETY: Never look directly at the Sun, either with the naked eye or through binoculars or a telescope; this could result in eye damage and even blindness.

Seeing the Moon

When the Moon is full, it looks quite bright, but don't be deceived! Full moonlight is only about one-millionth of the brightness of full sunlight.

The Moon is cold and rocky. It's a bit colder than the Earth so it doesn't give out its own light.

We see the Moon because it **reflects** sunlight towards us. Rays of light from the Sun bounce off the Moon's surface and travel into our eyes.

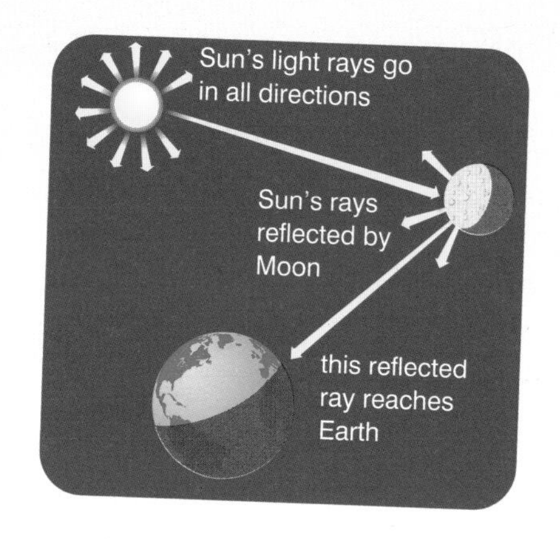
Sun's light rays go in all directions
Sun's rays reflected by Moon
this reflected ray reaches Earth

AMAZING SCIENCE!

The temperature of the Sun's surface is about 5500°C – that's hot! But it is much hotter inside, about 14 million degrees!

(You see this book in just the same way. Look around you. Light rays are falling on the book, and some of them reflect into your eyes.)

Q3 Draw a diagram to show how light from a lamp can help you read a book.

Gruesome science

We have a natural reflex which stops us looking directly at the Sun. Some illegal drugs stop this reflex from working, and people have gone blind as a result.

SUMMARY QUESTIONS

1 ☆ Copy and complete the sentences, using words from the list below.

light reflect star luminous

The Sun is a . . . . We see the Sun because it is a source of . . . .
The Moon and planets do not give out their own light; they are not . . . . We see them because they . . . light from the Sun.

Key words

luminous
reflects
star

7L2 Earth and Moon

LEARN ABOUT
- the Moon's orbit
- phases of the Moon

The Moon – our neighbour

Have you ever looked at the Moon through binoculars or a telescope? You might see mountain ranges and giant craters. You might also see flat areas. These are known as 'seas', although they are as dry as dust.

The Moon is cold. It does not give out its own light, so we describe it as **non-luminous**. We see it because it reflects rays of sunlight.

Moon watching

If you watch the Moon, you will notice that it rises in the east and sets in the west, just like the Sun. It follows a similar path across the sky.

Why does the Moon seem to move like this? Each day, the Earth spins from west to east. As a result the Moon *appears* to move from east to west.

In fact, the Moon travels around the Earth quite slowly. It takes about 28 days to complete one **orbit**.

We say that the Moon is the Earth's **satellite**. A satellite is a smaller object travelling around a larger object.

Q1 The Earth is a satellite. What object does it travel around?

Phases of the Moon

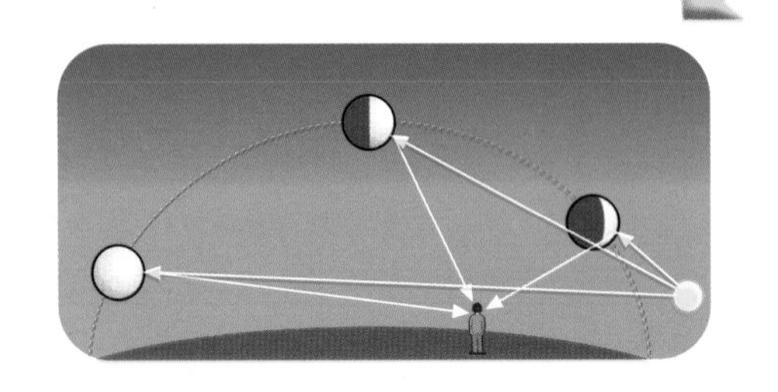

The shape of the Moon appears to change during a month. Sometimes we see a round, full Moon; sometimes we see a half Moon. Sometimes, when the Moon is 'new', we see a thin crescent Moon. These are the **phases** of the Moon.

We see the Moon because it reflects sunlight. When the Moon is opposite the Sun in our sky, we see a full Moon. When it lies in the same direction as the Sun, we only see the crescent-shaped edge of the side that is illuminated.

Bird's-eye view

We are used to seeing the Moon from the Earth. However, to understand our observations completely, we need to leave the Earth and look down on it from above. The picture shows what we could see from high above the North Pole.

You have to imagine standing on the Earth and looking towards the Moon. What would you see at each of its positions? The small pictures in the boxes show the views you would get as the Moon travels around the Earth.

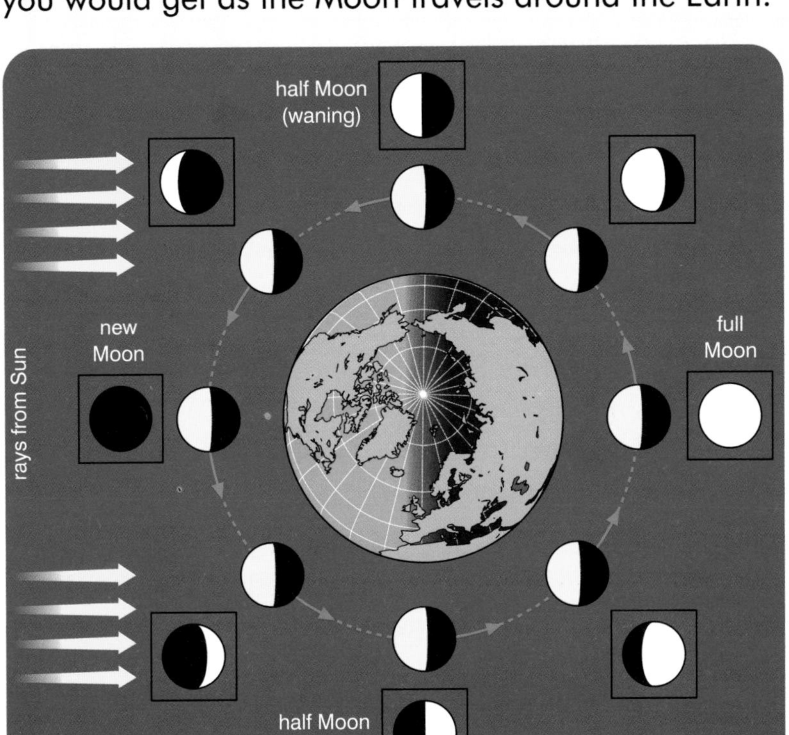

The Earth and Moon, seen from above. The Earth and Moon each have one side lit up by the Sun. The boxes show the phase of the Moon at each position around its orbit.

SUMMARY QUESTION

1 ☆ Copy and complete the sentences, using words from the list below.

orbit reflected phases satellite

The Moon is non-luminous. We see it by . . . light.

The Moon travels around the Earth. It is our . . . .

It takes about a month to travel once around its . . . .

Full Moon, half Moon, new Moon are all different . . . of the Moon.

Key words

non-luminous
orbit
phase
satellite

Four seasons or two?

LEARN ABOUT
- the Earth's tilted axis
- why we have seasons

Spring into summer

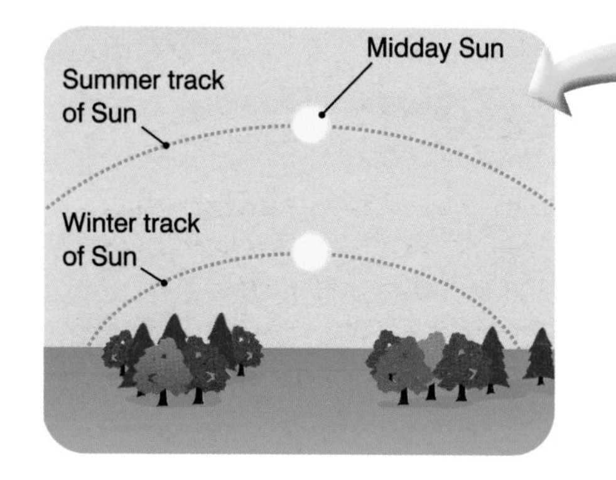

In the UK, we experience *four* **seasons** in the year – spring, summer, autumn and winter. In summer, the days are long and the Sun rises high in the sky. In winter, the days are much shorter and the mid-day Sun is low in the sky.

Not everyone experiences four seasons in the year. If you lived in the tropics, you would probably have *two* seasons – wet and dry.

Around the Sun

Why do we have seasons? The Earth moves around the Sun. It takes a year to complete one **orbit** of the Sun.

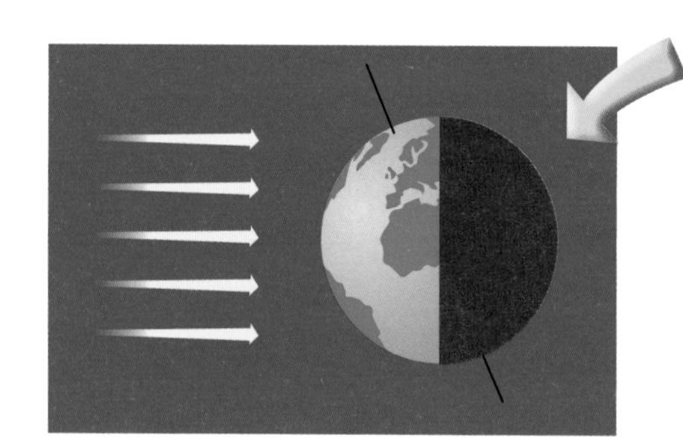

The Earth spins on its **axis** (the imaginary line through the poles). Its axis is *tilted*. During summer in the UK, our half of the Earth is **tilted** *towards* the Sun. The Sun rises high in the sky, and days are longer.

During the winter, the Earth has travelled around to the other side of its orbit. Now our half is tilted *away from* the Sun. The Sun stays low in the sky, and days are short.

Q1 Which part of the Earth is tilted towards the Sun in December?

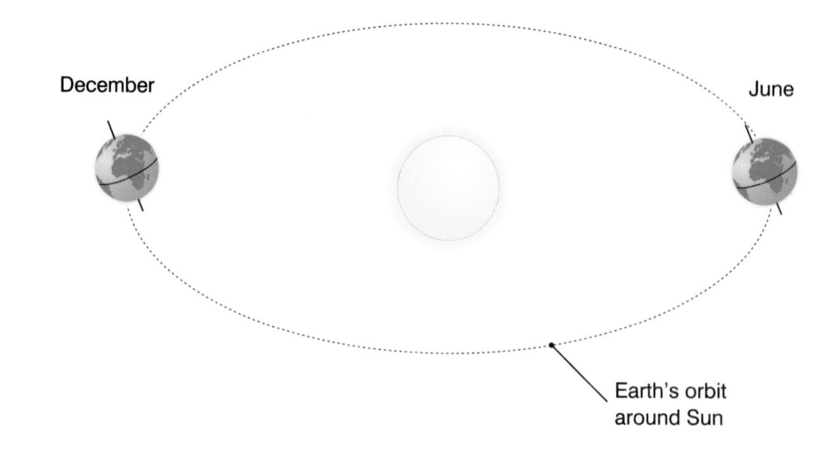

In winter, our half of the Earth is tilted away from the Sun. In summer, it is tilted towards the Sun.

Gruesome science

A year on Mars is longer than an Earth year, so its seasons are longer. In winter, the temperature at the poles can drop to −130°C – that's cold!

Sloping rays

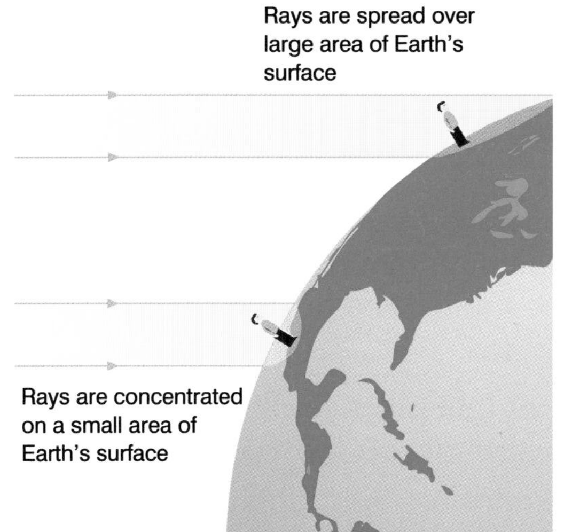

In northern latitudes, the Sun's rays are oblique. The person has a long shadow

In the tropics, the Sun's rays are almost vertical. The shadows are short

In summer, the Sun rises high in the sky. Its rays shine straight down on us, so they are concentrated on a small area. That makes it hot, and our shadows are short.

In winter, the mid-day Sun is low in the sky. Its rays are spread over a larger area, so it is colder and our shadows are longer.

Q2 Explain why your shadow is shorter at mid-day than in the evening.

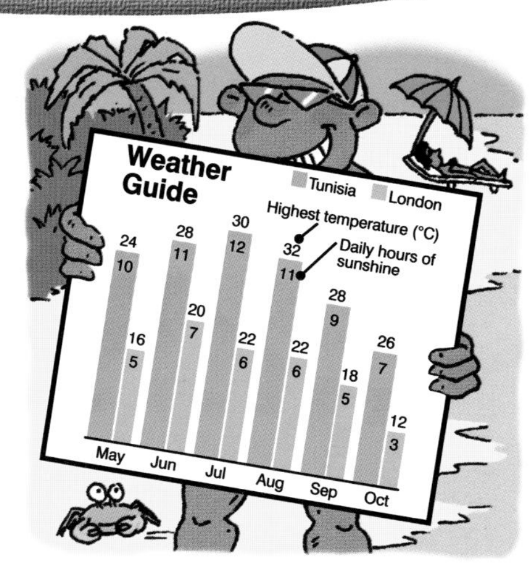

You can find graphs like this in holiday brochures. They compare the temperature and hours of sunshine at a holiday resort with those in London.

SUMMARY QUESTIONS

1 ☆☆ Copy and complete the sentences, using words from the list below.

summer axis winter orbit

The Earth travels in its . . . around the Sun. Its . . . is tilted.

In . . ., our part of the world is tilted towards the Sun. In . . ., it is tilted away from the Sun.

2 ☆☆ The diagram shows the path of the Sun across the sky. One path is for summer, the other for winter.

a) Which line shows its path in summer?

b) At what time of day is the Sun highest in the sky?

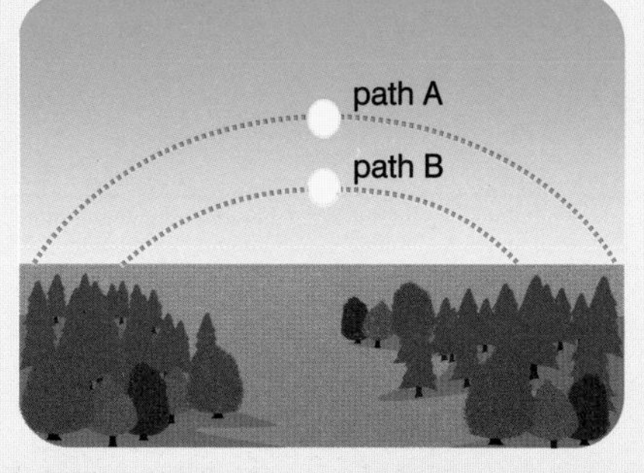

Key words

axis
orbit
seasons
tilted

Eclipses

LEARN ABOUT

- eclipses of the Sun and Moon

Eclipses of the Sun

Eclipses of the Sun occur quite frequently, up to three times a year. They happen when the Moon passes directly between the Sun and the Earth. The Sun's light is blocked off for a few minutes.

You have to be in just the right place on the Earth to see an eclipse of the Sun. This is what you see:

- At first, the Sun looks normal – a bright disc in the sky.
- Gradually, it becomes dimmer. If you look through special filters, you see that its surface has been blocked out.
- Eventually, the Sun is completely blocked out and the sky goes dark for just a few minutes.
- Then the Sun slowly returns to normal.

These young people are waiting for an eclipse of the Sun to happen. They are wearing glasses with special filters to protect their eyes from the Sun's rays.

Shadow of the Moon

The Moon has a **shadow** because it blocks out the Sun's rays. During an eclipse of the Sun, the shadow falls on the surface of the Earth. If you stand at this point, you will see a total eclipse of the Sun.

Q1 What do you think happens to the temperature during an eclipse of the Sun?

The Ancient Greeks saw that the Earth had a round shadow and realised that the Earth must be a sphere.

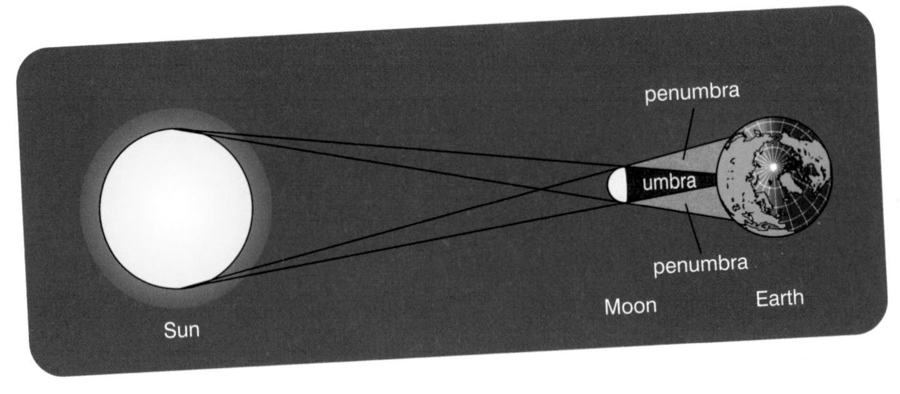

An eclipse of the Sun happens when the Moon blocks out the Sun's rays

Eclipses of the Moon

These photos show the Moon as it passes through the Earth's shadow during an eclipse

Every now and then – perhaps once or twice a year – there is an eclipse of the Moon. The full Moon gets gradually darker until it becomes very dim. Then it gradually lights up again.

This happens when the Moon's orbit takes it into the Earth's shadow, so that it is no longer lit up by the Sun. If you are anywhere on the dark side of the Earth, you will see an eclipse of the Moon.

Q2 Which has the biggest shadow: the Earth or the Moon?

Gruesome science

During an eclipse of the Sun in 1999, some people insisted that pregnant women should stay indoors because the were afraid that 'harmful rays' could affect an unborn child. (There were no harmful rays!)

Make your own eclipse

- Cut out two cardboard discs of the same size. One must be yellow or orange, the other black. These represent the Sun and the Moon.
- Use your discs to make your own eclipse. Show how the Sun and the Moon move across the sky. Then show what happens if the Moon passes in front of the Sun.

A lunar eclipse happens when the Moon passes through the Earth's shadow

SUMMARY QUESTION

1 ☆ Copy and complete the sentences, using words from the list below.

eclipse shadow Earth

The Moon is eclipsed when it travels into the Earth's . . . .

A total . . . of the Sun occurs when the Moon's shadow falls on the . . . .

Key words

eclipse of the Moon
eclipse of the Sun
shadow

Learn about

- the objects which make up the solar system
- what planets and their atmospheres are made of

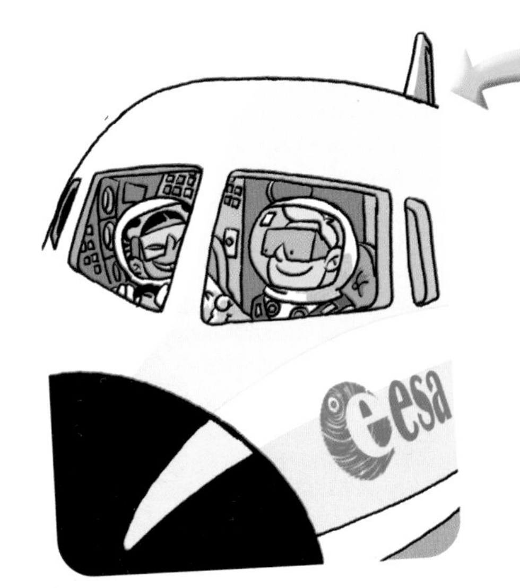

Journey into space

Strap yourself in, make sure the oxygen supply is working ready for launch . . . . We have lift-off! You're off on a journey of discovery, exploring the solar system.

At first, you are travelling upwards through the Earth's **atmosphere** – that's the layer of air around the Earth.

Beyond the Earth's atmosphere, you are out in space. Keep a look out for **planets**!

Head for the Sun

The closer you get to the Sun, the hotter it gets. The planets close to the Sun are hot as well.

Mercury is closest. Its average temperature is 120°C.

Venus is farther out, but hotter – a scalding 460°C! It has a dense atmosphere which holds the heat in.

Out past Mars

Mars is further than Earth from the Sun, so it is colder

Mars is the next planet beyond Earth as you travel away from the Sun. It looks red because its surface is covered with reddish sand. Mars is the last of the rocky planets.

Take care as you travel through the **asteroid** belt. An asteroid is a lump of rock in space, and many thousands of them lie in the asteroid belt between Mars and Jupiter.

Q1 How many rocky planets are there in the solar system? Make a list.

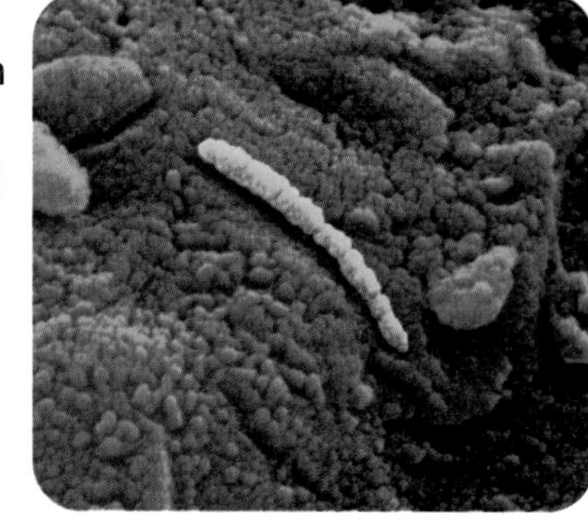

These blobby objects were found in a rock from Mars. Could they be fossilised bacteria? Does this prove that there was once life on Mars?

The gas giants

Jupiter and Saturn are two gas-giant planets. They are made mostly of liquid hydrogen, but each has a **core** of solid rock.

It's cold where these planets orbit, far from the Sun. That's why gases such as hydrogen and nitrogen are liquid or even solid there.

Beyond these gas giants are three more planets: Uranus, Neptune and Pluto. They are made of water, methane and carbon dioxide, all frozen solid.

ICT CHALLENGE

Choose one of the following subjects:

- planets
- moons
- asteroids

Make an electronic scrapbook of photographs, collected from the Internet or from CD-ROMs.

Q2 Why is there ice on Uranus, but no water?

Jupiter is famous for its Great Red Spot, a giant storm which has been raging in its atmosphere for hundreds of years

Many moons

The Earth isn't the only planet to have a **moon**. Other planets also have moons orbiting them. Saturn holds the record at present, with over 30 known moons. Astronomers use modern telescopes to search for new ones.

This is not a giant pizza. It's Io, one of Jupiter's moons. It orbits Jupiter in less than 2 days.

SUMMARY QUESTIONS

1 ☆ Name the Earth's natural satellite.

2 ☆☆ a) Name the four rocky planets.
b) Name two of the gas-giant planets.

3 ☆☆ Why would it be difficult for people to live:
a) on Mars?
b) on Jupiter?

Key words

asteroid
atmosphere
core
moon
planets

Learn about

- the Sun and other stars
- the night sky

Seeing stars

The Sun is a **star**. There are billions of billions of stars in the **Universe**. We can only see a few hundreds with the naked eye, but telescopes reveal many more. Each star is a glowing ball of hot gas, like the Sun.

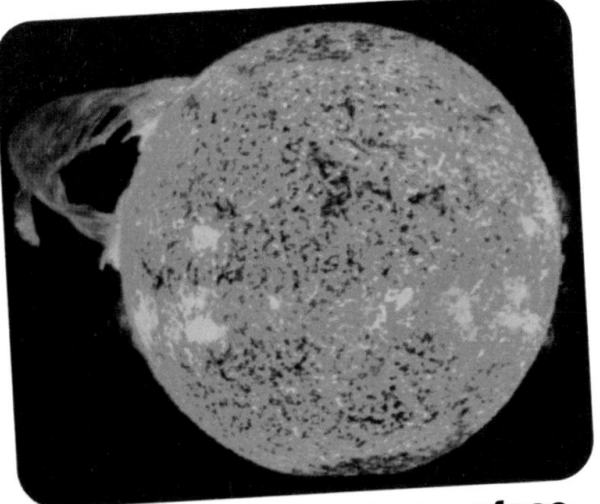

A close look at the hot surface of a star – the Sun

The Hubble space telescope took this picture of some of the most distant galaxies (clusters of stars) in the Universe. It shows them as they were formed, not long after the Universe came into existence.

At night, the stars are distant twinkling specks of light. If you step out of a brightly lit room, it will take a while for your eyes to adapt to the darkness so that you can see the stars.

During the daytime, our sky is brightly lit up and the stars are too faint to see. At dusk, the sunlight fades and the brightest stars begin to appear.

Don't forget the postcode!

Our galaxy is called the Milky Way.

- Imagine that someone wanted to send you a letter from a planet orbiting a star in a different galaxy. How would they address the envelope?

Q1 Explain why street lights make it difficult for us to see the stars.

Going around

It took $1\frac{1}{4}$ hours to take this photograph. The stars moved steadily across the night sky.

If you watch the stars at night, you may notice that their positions gradually change. They move across the sky at a steady rate, just as the Sun does during the day – and for the same reason. It's because the Earth is turning.

You may be familiar with the patterns of some **constellations** – groups of stars which look as if they are close together. People have navigated by the stars for thousands of years. Some birds fly at night, guided by the stars.

Q2 If the Earth turns from west to east, in which direction do the stars appear to move?

Gruesome science

In 1600, an Italian monk called Giordano Bruno was burned at the stake because he claimed that other planets, beyond the solar system, might be home to intelligent beings.

SUMMARY QUESTIONS

1 ☆ Copy and complete these sentences, using words from the list below.

night stars Earth planets Sun day

We see . . . because they reflect sunlight.

. . . produce their own light.

Light from the . . . makes it impossible to see the stars during the . . . .

The stars move across the sky at . . . because the . . . is rotating.

2 ☆☆ Find a map of the night sky, showing the current positions of the stars.
a) Which constellations are visible?
b) Why must you know the time to make use of the map?

Key words

constellation
star
Universe

 SCIENTIFIC PEOPLE 

Francisco Diego is an astronomer from Mexico. He travels the world to see **solar eclipses** and to study them. Francisco's son took one of these photos.

Francisco Diego has observed more than 10 solar eclipses

The Moon and the Sun look the same size in the sky. In fact, the Sun is 400 times as big as the Moon, but it is 400 times as far away. That's why the Moon exactly blocks out the Sun during a total eclipse.

When the Sun's bright disc is blocked by the Sun, we can see its corona. This glowing cloud of gas extends far out into space from its surface.

'To witness a total solar eclipse is a unique experience in your life.

'First, looking at the Sun through your safety filters, you see that more than half the Sun is covered. This is when you notice changes. The Sun is not hot enough any more. The landscape takes on a silvery appearance – something is really happening.

'Towards the west, the dark shadow of the Moon is rushing towards you. Now only a thin crescent of the Sun is left. The darkness increases minute by minute.

'Looking at the Sun, you see the magical Diamond Ring effect. When it is safe to remove your filters, you see the Sun's beautiful corona. You are inside the shadow of the Moon, and you see twilight around you in all directions. What you see, you will never forget – the solar corona, floating away from the Sun, all white, beautiful in the deep blue sky.'

Q1 Many people go on organised tours to watch eclipses of the Sun. Why do you think they go? Would you like to see an eclipse?

The stars seem to move across the sky because the Earth rotates once a day.

The Moon is smaller than the Sun.
It looks as big as the Sun because it is much closer.

If the Moon moves into the Earth's shadow, we see a lunar eclipse.
And if it moves in front of the Sun, we see a solar eclipse.

I can see Jupiter! We can see planets because they reflect the Sun's rays.

Imagine living on a planet orbiting one of those stars!
If its axis was tilted, you would experience seasons, just like on Earth.

- The Sun and other stars are luminous objects. They are sources of light.
- The planets, asteroids and comets all orbit the Sun.
- The Moon orbits the Earth – it is the Earth's satellite.
- We experience seasons because the Earth's axis is tilted.
- An eclipse of the Sun happens when the Moon moves in front of the Sun, blocking its light.
- An eclipse of the Moon happens when the Moon moves into the Earth's shadow.

DANGER! AVOID THESE COMMON ERRORS

It can be tricky to understand some of the scientific ideas about the solar system. To us, the Earth seems to be stationary, and the Sun and Moon seem to go around the Earth every day.

But think of seeing the Earth from a different viewpoint. Imagine you are floating in space, high above the North Pole. You will see the Earth turning as it **orbits** the Sun. You will see the Moon on its monthly orbit around the Earth. Perhaps you have seen an animation to help with this.

And don't forget the tilt of the Earth's **axis**. It stays tilted in the same direction. It doesn't tilt back and forth during the year.

Key words

axis
orbit
solar eclipse

UNIT REVIEW

REVIEW QUESTIONS

Understanding and applying concepts

1 We experience seasons because the Earth's axis is tilted. But there is another reason why a planet might have seasons. The Earth's orbit around the Sun is almost circular. Another planet might have an elliptical orbit, like the one in the picture.

- a Explain why this would lead to seasons on the planet.
- b How would these seasons be different from our own?

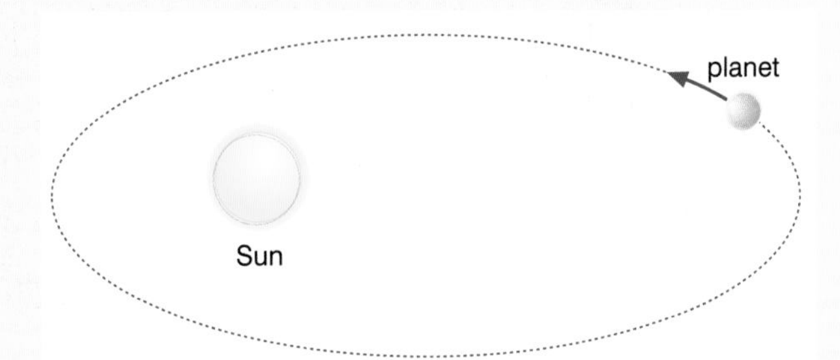

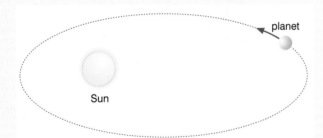

Ways with words

2 A solar eclipse happens when the Moon blocks the Sun's rays. Write another sentence, nothing to do with eclipses, using the word 'solar'.

3 A *mnemonic* is a way of remembering something. Make up a mnemonic which will help you to recall the names of the planets in the correct order. (The order is: Mercury, Venus, Earth, Mars, Jupiter, Saturn, Uranus, Neptune, Pluto.)

Making more of maths

4 Follow the instructions which follow to draw a scale diagram of the Earth and the Moon. You will need a big sheet of paper.

- Represent the Earth by a circle 12 mm in diameter.
- Represent the Moon by a circle 3 mm in diameter, 36 cm from the Earth.
- Shade half the Earth to show where it is night. Shade half the Moon in the same way.
- Imagine that you are able to look up at the sky and see the Moon. Use your drawing to help you show where on the Earth you might be.

On which part of the Earth would you not be able to see the Moon?

Thinking skills

5 Spacecraft have visited Mars and looked for signs of life. They have found nothing. One of your friends says that this proves that there is definitely no life on Mars. Do you agree?

Extension question

6 This unusual photo was taken from the *Voyager 1* spacecraft. It shows both the Earth and the Moon.

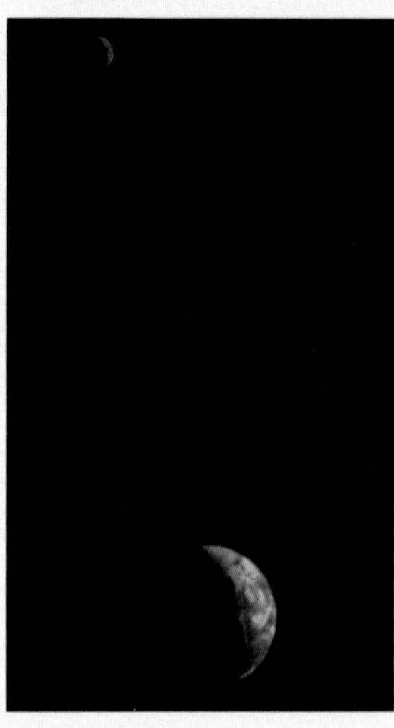

- a Imagine looking down onto the Earth and Moon. Draw a picture to show the relative positions of the Earth, Moon and Sun at the time this photo was taken.
- b Show the parts of the sides of the Earth and Moon that are lit up by the Sun.
- c Explain how your picture relates to the photo. Why do both Earth and Moon look like crescents?

SAT-STYLE QUESTIONS

1 Every day, the Sun appears to travel across the sky. The picture shows the path of the Sun across the sky during a winter's day.

a The picture shows the position of the Sun at mid-day. Explain how you can tell from the picture that it is mid-day. (1)

b Copy the picture and add arrows to show how the Sun moves along its path from dawn to dusk. (1)

c Draw another line on the picture to show the path of the Sun across the sky on a summer's day. (1)

2 The picture shows the Earth with the Sun's rays shining on it. Answer the following questions, giving reasons for your answers.

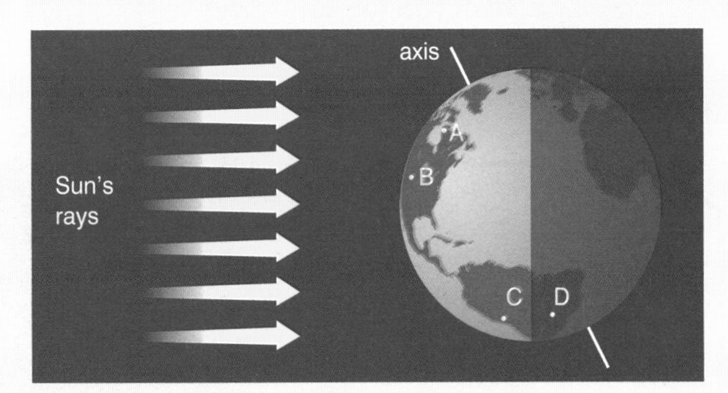

a At which point is it night-time? (2)

b At which point is the Sun highest in the sky? (2)

c At which points is it winter? (3)

3 During 2003, an eclipse of the Sun could be seen from the north of Scotland. Paul wore special safety glasses to protect his eyes when he watched the eclipse.

Paul says: 'You must wear safety glasses because the Sun is a luminous object. You don't need to wear them for an eclipse of the Moon, because the Moon is a non-luminous object.'

a Paul says that the Sun is a luminous object. Is he correct? (1)

b Paul is wrong when he says that safety glasses must be worn when looking at the Sun because it is luminous. Give the correct reason. (1)

c The picture shows the positions of the Sun, Moon and Earth a few days before the eclipse. Copy the picture. Add to the picture to show how light from the Sun allows us to see the Moon. (2)

Key words

Unscramble these:

paces
rats
talpen
verisune

Back in Year 6, Molly and Benson investigated dissolving. You may have carried out a similar investigation. Look at what Molly and Benson did – do you think they made the most of their investigation?

Decide on the question you are going to investigate.

Which dissolves quicker, salt or sugar?

Can hot water dissolve more stuff, and does it dissolve things quicker?

- Think of a question that you can test.

So the question is: *Does salt dissolve more quickly in hot water than in cold water?* Now try to make a prediction.

table salt

It will dissolve faster if you use hot water.

How did you know that?

- Use what you already know.

Collect your equipment. I'll bring round the hot water.
More tea, vicar?

- Make the most of your equipment by making accurate measurements.

How long did that take?
I don't know, but it's 4.30 in Karachi.

- A stopwatch is more precise than the clock on the wall.

- Think about how best to present your results.
- A suitable graph can help you make the most of your results.

- Think about how you did your investigation – evaluate it.

- Use your scientific knowledge to explain what you have found out.

Key words

accurate
evaluate
pattern
precise
prediction
results

During their science studies at Scientifica High, Benson and Molly will learn a lot about how to carry out investigations. Look back at the two previous pages, and think about how they might have improved their investigation.

ACTIVITY
Investigating dissolving

Molly and Benson investigated dissolving. There are lots of questions about dissolving which they could have investigated. Their question was:

Does salt dissolve more quickly in hot water than in cold water?

- Suggest another question about dissolving which Benson and Molly might have investigated.

Scientists try to think ahead. They make a **prediction**, saying how they think their investigation may turn out.

- What prediction did Benson and Molly make?
- What **evidence** can you provide to support their prediction?

(Remember, evidence can come from anywhere – from your scientific knowledge and from everyday life.)

Benson and Molly weren't very happy with their equipment. They thought it might not give them very **accurate** measurements.

- Look at the equipment they used. How could it have been improved, to give them more accurate measurements?
- What should they have done to make sure their investigation was a fair test?

Scientists try to use the best equipment, but they also try to use it as carefully as possible, to ensure that their results are as accurate as possible.

- How could Benson and Molly have made their results more reliable?
- Benson and Molly had cold water from one tap, and hot water from a jug. They mixed them to make warm water. How could they have improved their investigation by including higher and lower temperatures?

Benson and Molly made a mess of presenting their results.

- Show how they could have written down their results, so that the **pattern** was easier to see.
- Other groups in the class investigated the same question. Benson and Molly combined their results with the results from the other groups. Why was this a good idea? Why might it lead to problems?
- What type of graph would have been a good way to show their results? Draw the graph.
- Write a sentence to describe the pattern in the results.
- Can you use ideas about particles to explain the pattern?

Pete and Reese are investigating burning. Look at how they carried out their investigation – are they better at investigating than when they were in Year 6?

On page 215, Pip and Mike are investigating how the temperature changes during the day. They are using electronic equipment to collect data, but they still have to think just as hard about what their results mean.

What is the question you are going to investigate?
Will a candle burn longer if it has more air?
A candle needs oxygen from the air.
But perhaps it will be completely burnt up before it has used all the air.

- Think ahead – can you imagine the pattern of your results? Use your scientific knowledge.

When should I start the clock?

As soon as the beaker touches the sand.

- It can help to have a trial run.

It's tricky to judge just when the flame goes out.

Better just write 28 seconds, then.

- Repeat readings can help you to judge how reliable your measurements are.

How good is your evidence?
We've only got values for two sizes of beaker.
size time
We needed some more beakers to be sure of the pattern... and we needed more time!

- Think about the strength of the evidence you have collected.

Key words

accurate
evidence
pattern
prediction
reliable

Best practice in Sc1

- A graph will help you test your prediction.

Here's Pete and Reese's data. Where would you draw a line of best fit through the points?

Time

We had just two points on our graph.

A perfect straight line!

- A line of best fit has the points scattered on either side of it.

ACTIVITY
Better, stronger evidence

In a scientific investigation, you have to make the most of what you've got – equipment, materials and ideas. Pete and Reese did their best, but they needed more time.

They had to measure the time for which the candle burned.

- Why was it a good idea to have a trial run?
- Why was it a good idea to make **repeat readings**? How did this help them to decide whether their results were reliable?

The teacher showed the graph of Pete and Reese's results. A graph can also help you to decide whether your results are **reliable**.

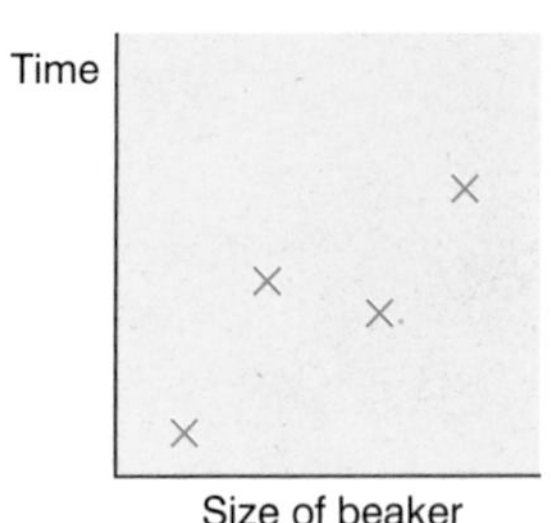

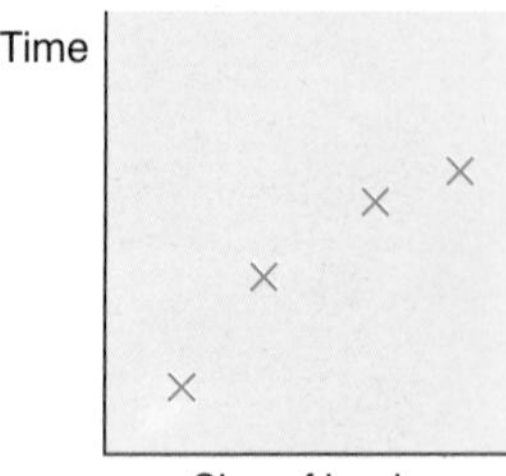

- Look at the two graphs shown here. Do they both show the same pattern? How does a graph help you to decide whether you results are reliable?
- How does a **line of best fit** help to show the pattern in some results? How could it help you to make predictions?

- An electronic data logger can help you in investigations.

- A computer can handle lots of data very quickly.

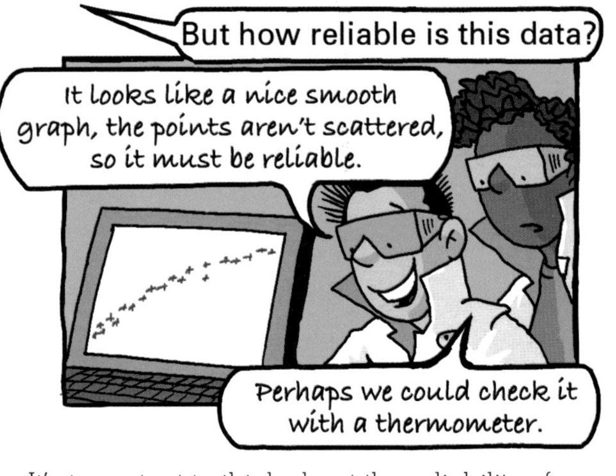

- It's important to think about the reliability of data gathered by sensors.

- You can share data through the internet.

ACTIVITY Electronic data

An electronic data logger, equipped with sensors, can help in an investigation.

- Explain why a data logger was useful in this experiment.

Data loggers collect data automatically. They can do it over a long period of time. They can also collect data very quickly.

- Suggest an experiment where a data logger would be useful to collect data quickly.

If you use sensors and a data logger, you need to check their reliability.

- How could Pip and Mike have used a thermometer to assess the reliability of the data from the data logger?

The class had some secondary data, from Australia.

- How could this help the class to understand the pattern in their own data?

Key words

data logger
line of best fit
reliability
repeat readings
sensors

7A Cells

7A1 Use a microscope to examine a very small specimen provided by your teacher. Make an accurate drawing of what you see.

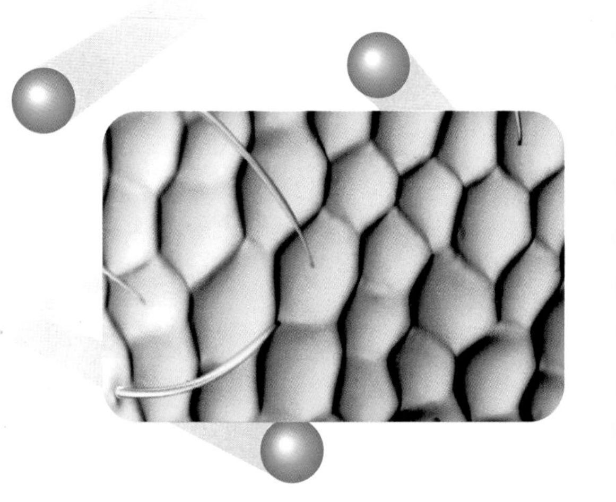

7A2 Paint the underside of a leaf with a layer of pale pink nail varnish to make a 'cast' of the leaf surface. Use your microscope to look at your 'cast' of the leaf surface.

a) What shape are the cells that make up the surface?

b) Can you see hairs on the leaf?

You should be able to see holes, called *stomata*, which allow air into a leaf.

7A3 Make a hay infusion from some grass or hay and a small amount of water, from a pond if possible. Use a binocular microscope to investigate the microscopic pond life that develops.

7A4 One of your most important organs is your liver. Sometimes it is called the body's chemical factory. Find out what your liver does.

7A5 It is less than 100 years since we really began to understand what cells are and how we make more. Find out about the important discoveries made by the following people:

- Theodor Schwann (1810–1882), a German zoologist
- Oscar Hertwig (1849–1922).

7B Reproduction

7B1 Compose an Agony Aunt page for a magazine aimed at teenagers.

a) Write some letters that ask about the problems of going through puberty and adolescence.

b) Write short, but sensible, replies to the letters.

7B2 Your chromosomes determine whether you are a boy or a girl.

Find out how you inherit your sex.

7B5 Some mothers cannot breast-feed their babies.

a) Work out how much it costs to buy the milk to bottle-feed a baby for 6 months. You will need to find out:

i) how much a tin of milk costs (try a supermarket web site)

ii) how many feeds can be made from a tin

iii) how many feeds a baby needs each day.

b) What else will the new mum have to buy to prepare for bottle-feeding?

7B6 Birth and the first year are the riskiest times of life. Millions of babies and children in developing countries die from drinking water. Their water contains harmful bacteria that give them diarrhoea. The best treatment for infected children is oral rehydration salts.

a) Find out what is in oral rehydration salts.

b) Your parents probably gave you a home-made version of the salts when you were a small baby, and had a tummy bug that made you be sick.
What did they use?

Key words

Research these new words:
chlorophyll
chromosome
fertility drug
protozoa

7C Environment and feeding relationships

7C3 Play the Yes/No game. Think of an animal that is nocturnal or adapted for life at night. The others in your group can ask up to 10 questions to try and identify it. They can only ask questions with a 'Yes' or 'No' answer. What are the most useful questions to ask?

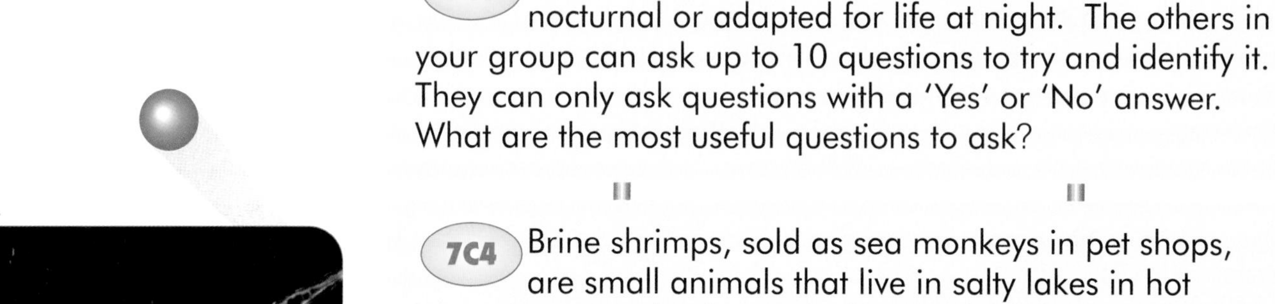

7C4 Brine shrimps, sold as sea monkeys in pet shops, are small animals that live in salty lakes in hot countries. They feed on tiny algae in the lakes. Brine shrimps are adapted to survive conditions that change from very salty water to totally dried out.

a) Find out more about how brine shrimps are adapted.

If you are lucky enough to have brine shrimps at school you can see that they seem to swim upside down. Find out what happens if you change the direction of light from above their container to below it.

b) Are they sensitive to the direction of light?

c) How would this help them to survive better?

7C5 **a)** What is the world's biggest eater? It eats the smallest of food sources – krill.

b) Find out what krill is and what eats it.

c) Use what you have found out to make a food chain.

7C6 Predators catch, kill and eat their prey. Parasites need their prey to stay alive and provide them with lunch all their lives.

Find out about a parasite and how its body and way of life is adapted for its diet. You could find out about mosquitoes, leeches, bed bugs, head lice or cat fleas.

7D Variation and classification

7D1 During the last 300 million years many species have become extinct. We know about these from fossils. Find out about how fossils are made and what sorts of things can become fossilised.

7D2 Find out about crocodiles and alligators, or the octopus.

7D3 Plants do not move from place to place but they do move. For example a daisy's flowers open in the morning and close at night. The leaves of the sensitive plant droop when they are touched.
Find out what these plant movements are called and describe some more examples of plant movement.

7D4 Many of our features are inherited.
Do a survey in the class of the distribution of inherited features, such as a small gap between the two front teeth, or free or fixed earlobes. (If you have fixed earlobes your earlobe goes smoothly into the jaw and you cannot fasten clip-on earrings onto them very well. Free earlobes have a small or large rounded flap at the bottom.)

7D5 Find out more about DNA.

7D6 Many modern varieties of plants are described as hybrid, but pedigree dogs are pure bred.
Find out what 'hybrid' and 'pure bred' mean.

Key words

Research these new words:
decomposer
deforestation
extinction
hybrid
mutation

Phenomenal perfo

7E Acids and alkalis

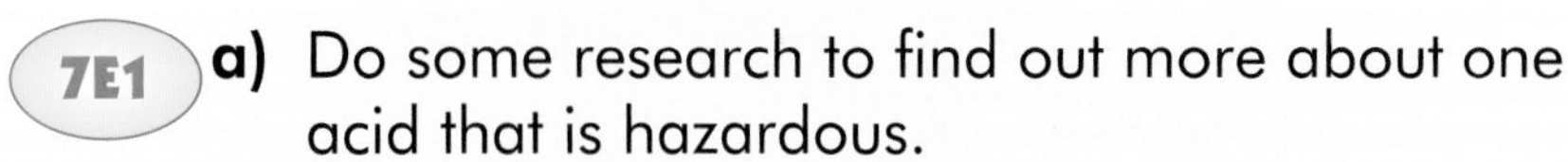

7E1 **a)** Do some research to find out more about one acid that is hazardous.

b) Present your findings as a poster to share with the rest of your class.

7E2 Find out how the emergency services would deal with a spill from this tanker lorry.

7E3 **a)** What makes a dye a good indicator?

b) Plan an investigation to find out which of three plants makes the best indicator.

7E4 When an acid and an alkali react together we get a substance called a salt, plus water formed. A salt is a solid substance made of crystals.

a) Why do you think that you didn't see a salt forming in your experiments?

b) How could you get a sample of a salt from the neutral solution formed?

7E5 Predict the change of pH you will get as you add sodium hydroxide to hydrochloric acid. You start with 25 cm^3 of acid, and the acid and alkali both have the same concentration.

7F Simple chemical reactions

7F1 **a)** Make a list of chemical reactions that take place in everyday life.

b) On mixing two solutions in a beaker, the temperature increased. What does this tell us about the change taking place?

7F2 Make a fact sheet on hydrogen gas for another Year 7 class.

7F3 Find out about different types of fire extinguisher and when they are used.

7F4 When magnesium burns in air it forms magnesium oxide.

Do you think the mass of reactants is the same, lower or higher when compared with the mass of product formed? Explain your answer.

7F5 Methane is made up of carbon and hydrogen. When methane burns, it produces carbon dioxide and water in its combustion reaction.

We can show this by a word equation:

methane + oxygen → carbon dioxide + water

Butane also contains carbon and hydrogen.

Write a word equation to show the combustion of butane in plenty of air.

Research these new words:
ammonia
buffer solution
hazcards
oxidation

7G The particle model

7G1 Make a list of materials that would be tricky to classify as a solid, a liquid or a gas. Explain your choices.

7G2 Democritus explained the properties of materials by saying that their particles were different. For example, a runny liquid must be made up of smooth, round particles so that they can tumble over each other. On the other hand, hard solids must be made up of particles that are sharp and jagged. These particles get stuck in position, so that explains why they don't flow and are hard.

a) Draw diagrams that show the ideas of Democritus described above.

b) Do you think his ideas made sense? Why?

7G3 Find out more about the life and work of either John Dalton or Robert Brown. Present your information as a brief radio report as part of a programme for schools.

7G4 Explain the properties of solids, liquids and gases (shown in the table on page 112) using the particle theory.

7H Solutions

7H1 A salt company mined rock salt from two different places, A and B. In place A, 250 tonnes of rock salt produced 150 tonnes of pure salt. In place B, the company can get 90 tonnes of pure salt from 120 tonnes of rock salt. Which place would be best to mine? Show your working out.

7H2 Sand is insoluble in water.
Use the particle theory to try to explain why.

7H3 Imagine you are stranded on an isolated island in the middle of the Pacific Ocean. Your only fresh water is in a stream, but the water looks dirty.
Describe how you could get pure water from the stream.

7H4 Draw a flow diagram that describes how to set up a chromatogram.

7H5 Use your graph from Question 2 on page 135 to answer the question below:
How much potassium sulphate will dissolve in 25 g of water at 60°C?

Key words
Research these new words:
benzene
crystal gardens
plasma

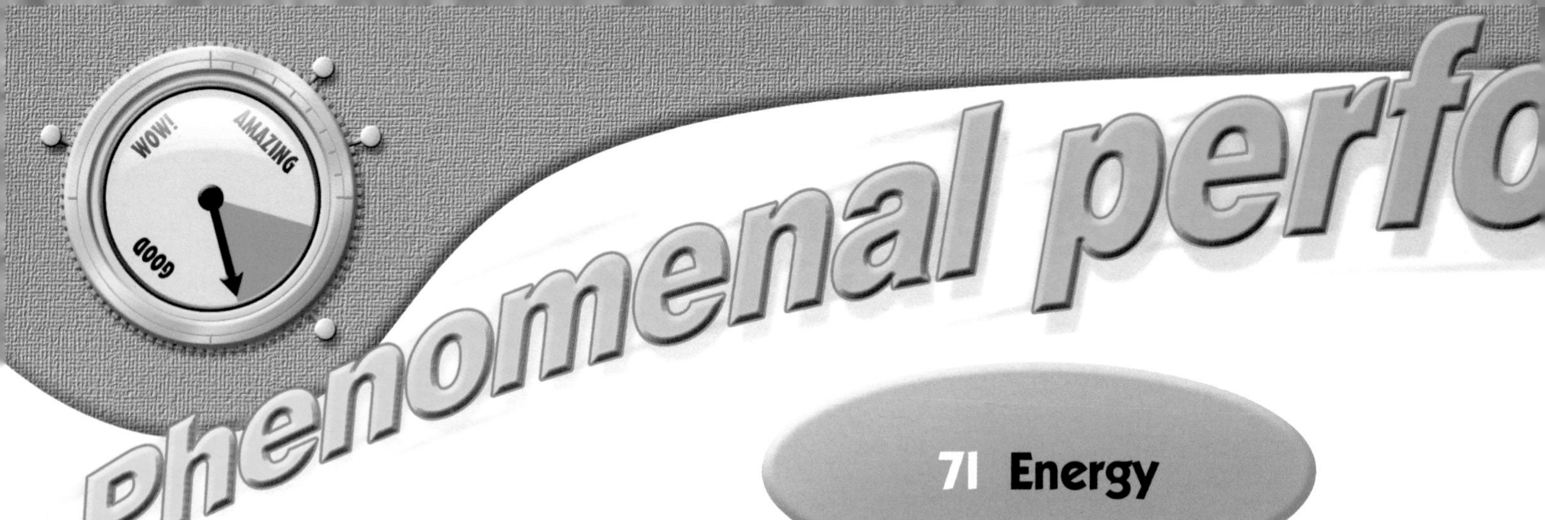

71 Energy

711 Imagine a day in which you use lots of different fuels. Make up a diary for your imaginary day, like this:

Activity	Fuel used
Toast and tea for breakfast	Gas
Went to school on the bus	

712 Most of the energy we use comes from fossil fuels, but we use a lot of electricity in our homes.
Electricity brings us the energy of fossil fuels which are being burned in power stations.
Suggest as many reasons as you can why we often prefer to use electricity at home, rather than burning fossil fuels.

713 In History, you have probably studied how people lived before electricity came along.

a) What fuels did people use?
b) What did they use them for?
c) Were any of their energy supplies renewable?

714 School buildings use a lot of energy, and much of it is wasted.
Think about your school.

a) Can you think of any ways in which energy is wasted?
b) How could the school waste less energy? You can't knock it down and rebuild it, but there are probably other ways of altering it to save energy resources. Your headteacher may be glad to know of ways to cut the fuel bills!

7J Electrical circuits

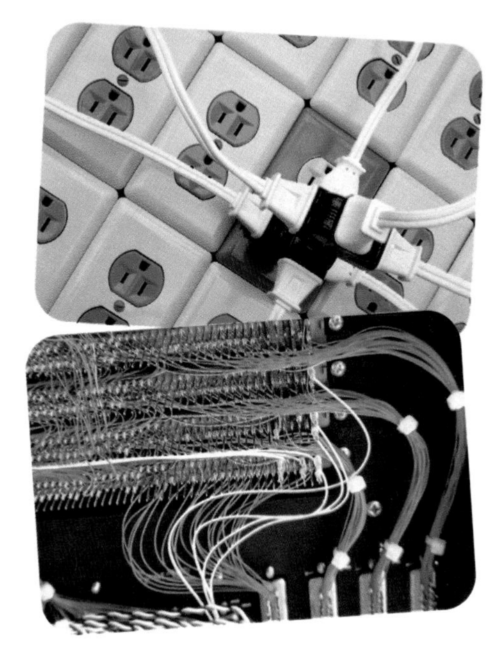

7J1 When you connect something to the mains, you use a flex with a plug on the end of it. Look at a length of flex.

a) How many wires does it have in it?
b) What materials is it made of?
c) What colours of insulation are used for the different wires?
d) Why is it important that the wires are different colours?

7J3 A car battery gives 12 V. It is made of several 2 V cells. Draw a circuit symbol to represent this.
Find some battery-operated devices, such as a mobile phone, laptop computer, calculator, watch, camera.

a) What voltage do they require? (You may have to look in the instruction book.)
b) Do they have a single cell or a battery of several cells?

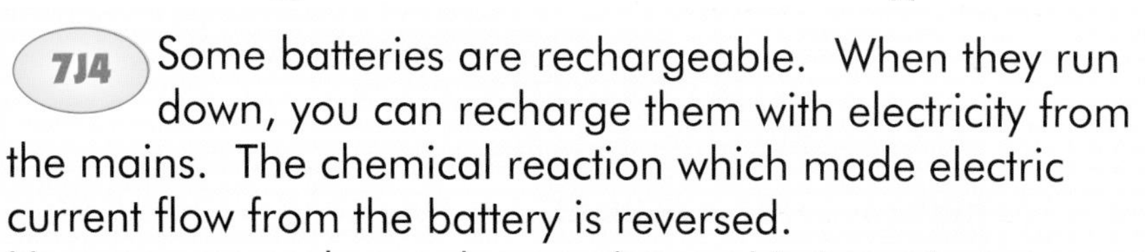

7J4 Some batteries are rechargeable. When they run down, you can recharge them with electricity from the mains. The chemical reaction which made electric current flow from the battery is reversed.
Here are some places where rechargeable batteries are used: invalid cars, mobile phones, milk floats, laptop computers.
a) Can you think of any others?
People don't use rechargeable batteries as much as they might. They prefer to use ordinary batteries and throw them away when they are flat.
b) Why do you think this is?

Key words
Research these new words:
earth wire
fuel cell
nuclear power
RCD

7K Forces and their effects

7K2 Here's a way to make a weighing machine:

- Take a long wooden stick, such as a metre rule from the science lab.
- Clamp one end to the bench, so that the other end sticks out.
- Hang a heavy object on the end, so that the stick bends downwards.

a) Draw a diagram of this.
b) Explain how you could turn this into a weighing machine.

7K3 If you go to the Moon, your mass stays the same but your weight is a lot less. People think that there will be holiday trips to the Moon one day. The photo shows an idea for a Moon hotel.

Design a brochure to encourage people to come and stay in the hotel. Describe the facilities and any activities for visitors.

7K6 The Eurostar train travels from London to Paris. It goes through the Channel Tunnel.

For 30 minutes, the train travels slowly through London. Then it speeds up, and travels at top speed for 2 hours until it arrives in Paris.

a) Draw a distance–time graph to show the complete journey.
b) Draw another graph to show the return journey.

rmance

7L The solar system – and beyond

7L1 We have night and day because the Earth spins on its axis. Describe the changes you would expect to see if:

a) the Earth turned more quickly
b) the Earth turned more slowly
c) the Earth turned in the opposite direction.

7L3 You have learned the scientific explanation of why we have seasons. But, in the past, people had different explanations.
Try to forget what you have learned, and think up as many alternative explanations as you can for the seasons.

a) Why is it hotter at some times of the year than others?
b) Then think of evidence for and against your explanations.

7L6 Many astronomers complain that it is difficult to see the stars in the night sky. The glare from street lights is a kind of 'light pollution'.

a) Can you get a clear view of the stars from where you live?
b) Have you had a better view when you have been on holiday?
c) Do you think it is important for ordinary people to be able to see the stars at night?

Key words
Research these new words:
annular eclipse
space telescope
neutron star
supernova

GLOSSARY

acid — a substance that forms a solution with a pH of less than 7. p.80

adapted — being specialised for a particular activity, or particular conditions. pp.8, 42

air resistance — similar to drag; the force of friction when an object moves through air. p.185

alkali — a substance that forms a solution with a pH greater than 7. p.84

ammeter — an instrument for measuring electric current. p.160

amniotic fluid — the fluid that cushions a growing foetus. p.28

asteroid — a rock in orbit around the Sun. p.202

atmosphere — the layer of gas surrounding any planet. p.202

axis — the imaginary line through the middle of the Earth, about which it turns. p.198

balanced forces — two or more forces whose effects cancel out. p.182

battery — two or more electrical cells connected together. p.162

biomass — living material that can be burned to release energy. p.146

calcium carbonate — the main substance in limestone, chalk and marble. Pure calcium carbonate is a white solid. It fizzes in dilute acid releasing carbon dioxide gas. p.101

carbon — a chemical element that exists as diamond and graphite. It is also found in charcoal and soot. p.104

carbon dioxide — a gas that makes up about 0.04% of the air. It is produced when we burn fossil fuels. p.100

cell — the basic units living things are made of. p.6

cell a single component that provides a voltage in an electric circuit. p.158

cell division how cells make new cells. p.12

cell membrane the outer layer of a cell. It controls what enters and leaves a cell. p.6

cell wall a tough outer layer around plant cells that gives them strength and shape. p.7

chlorophyll green substance that catches light energy in plants. p.62

chloroplast plant cells. They make food by photosynthesis. p.7

chromatogram the paper that shows the separated substances after chromatography. p.132

chromatography the process whereby small amounts of dissolved substances are separated by running a solvent along a material such as absorbent paper. p.132

chromosome found in the nucleus of a cell. It carries genes. pp.12, 70

classification arranging organisms in groups with similar features. p.63

circuit diagram a way of showing how the components of a circuit are connected together. p.158

comet a lump of ice and dust in orbit around the Sun. p.203

competition when two animals or plants both need the same resource. Each finds it harder to get what they need. p.54

conserving energy resources making careful use of energy resources, so that they don't get used up too quickly. p.149

consumer an animal that eats other animals or plants. p.50

corrosive describes substances that attack and destroy living tissues, including eyes and skin. p.82

cytoplasm a jelly-like substance inside other cells. p.6

density calculated using density = mass/volume. p.183

Key words
on key people
Charles Darwin
variation and environment

diffusion the mixing and movement of one substance through another (without the need to stir up the substances). **p.118**

distillation the process of separating a pure liquid from a mixture. First boil the mixture containing the liquid. Then condense the gas and collect the pure liquid. **p.130**

dormant plants survive the winter by becoming dormant. They are alive but do not grow until spring. **p.49**

drag the force of friction when an object moves through a liquid or a gas. **p.185**

eclipse when the Sun is hidden because it passes behind the Moon, or when the Moon passes into the Earth's shadow. **pp.200–1**

electric circuit a complete path around which electric current can flow. **p.158**

energy resources anything from which we can get energy. **p.146.**

energy the ability to make things happen. For example, burning fuels release energy, allowing us to do things. **p.142**

equation a way of describing a chemical reaction that shows what we start with and what we finish with after the reaction. **p.103**

evidence data (measurements and observations) collected by scientists that support or challenge a theory/ prediction/conclusion. **p.114**

external fertilisation sperm fertilise eggs outside the body. **p.35**

fertilisation occurs when a sperm nucleus passes into the egg carrying its genes. **p.25**

filtration separating insoluble solid from a mixture of the solid and a liquid. You can do this by passing it through filter paper. p.127

foetus a baby developing in the uterus. It is not called a baby until it has been successfully delivered. p.28

food web a set of linked food chains in a habitat. p.50

force a push or pull which acts on one object, caused by another. p.174

forcemeter a device for measuring forces. p.176

fossil fuel a fuel formed from materials that were once alive. p.144

friction a force which opposes movement. p.184

fuel a material burned to release energy. p.142

fuse a component that melts ('blows') when the current flowing through it is too great. p.168

gas pressure the force per unit area exerted by gas particles as they collide with the walls of their container or other objects. p.119

gravity the pulling effect of any object that has mass. p.178

habitat the place where an animal or plant lives and reproduces. p.44

harmful describes substances that cause some damage to the body. They might be swallowed, breathed in or absorbed through the skin. p.82

hazardous potentially dangerous substances. p.82

hibernation spending tough winter conditions in a secure, sheltered place. p.48

hormones chemicals made by glands that pass round the body in blood. They control many activities. p.23

Key words
on key people
Leeuwenhoek microscopes and bacteria
Democritus particles

humidity the amount of moisture in the air. p.45

hydrochloric acid a common acid found in school laboratories. p.84

hydrogen a gas that is the lightest of all the chemical elements. A lighted splint burns with a squeaky pop when applied to the mouth of a test tube of the gas. p.99

in parallel components connected side-by-side are in parallel. p.166

in series components connected end-to-end are in series. p.166

indicator a substance that changes colour depending on the pH of the solution it is in. p.84

insoluble does not dissolve in a particular solvent. p.126

insulation material that prevents you from accidentally touching electricity, because it doesn't conduct. p.159

invertebrates eight very different groups of animals. None have a backbone. p.64

joule (J) the scientific unit of energy. p.150

kilogram (kg) the scientific unit of mass. p.180

limewater solution which turns milky/cloudy when carbon dioxide is bubbled through it. p.100

luminous describes an object that is a source of its own light. p.194

magnesium a grey metal (silver beneath its outer coating) that burns in air with a bright, white light. p.103

magnification the number of times an image is enlarged by lenses. p.4

mass how much matter an object is made of. p.180

methane the main substance in natural gas. It contains the elements carbon and hydrogen. p.104

migration animals moving to more favourable areas when conditions become harsh. p.48

model a 'picture' constructed by scientists to help explain the way the things work. p.114

newton (N) the scientific unit of force. p.176

nocturnal animals only active when it is dark. p.46

non-luminous describes an object that we can only see by reflected light. p.197

non-renewable energy resources energy resources that get used up and which are not replaced. p.146

nucleus found in cells. It controls the cell's activity and carries inherited information. p.6

orbit the track followed by one object as it travels round another object. p.203

organs structures that carry out jobs necessary for life. For example lungs obtain oxygen from the air. They have components that do different parts of the job. p.10

ovulation when an egg is released from the ovary. p.26

ovule found in flowers. It forms a seed after it has been fertilised by pollen. pp.14–15

oxygen the gas that is essential for breathing and burning. It makes up 21% of the air. A glowing splint re-lights in the gas. pp.102–3

palisade cell a cell in the upper layers of a leaf. It is specialised for photosynthesis. p.8

particle a tiny, individual piece of a substance that is too small to be seen. p.115

pH value a number on the pH scale that indicates how acidic or alkaline a solution is; a pH value of 7 is neutral. Values below 7 are acidic (the lower the number the more acidic a solution is). pH values above 7 are alkaline (the higher the value the more alkaline a solution is). p.85

Key words

on key people
Robert Brown
particle motion
Antoine Lavoisier
burning

pollen — found in flowers. It fertilises ovules. p.14

producer — plants are producers because they use light energy to make foods. p.50

product — a substance that is formed in a chemical reaction. p.97

puberty — the time when our reproductive organs mature. p.22

react — when substances undergo a chemical change and form new substances. p.96

reactant — a substance that we start with before a chemical reaction takes place. p.97

renewable energy resources — energy resources that will never run out. p.146

resistance — how much a component opposes the flow of electric current. p.161

rubella — a virus that harms foetuses. p.32

satellite — a smaller object travelling around a larger object. p.196

saturated solution — a solution in which no more solute will dissolve at that particular temperature. p.134

section — a thin slice of material for examining under a microscope. p.5

selective breeding — animals or plants with the best features are kept to be the parents of the next generation. Gradually desirable features are spread through the population. p.72

sodium hydroxide — the most common alkali found in school laboratories. p.84

soluble — describes a substance that dissolves in a particular solvent. p.126

solute — a substance that dissolves in a particular solvent. p.126

solution — the mixture formed when a substance dissolves in a liquid. p.126